ENTERING SPACE

United States
USA

JOSEPH P. ALLEN
WITH RUSSELL MARTIN

ENTERING SPACE

AN ASTRONAUT'S ODYSSEY

DESIGN BY HANS TEENSMA

ORBIS.LONDON

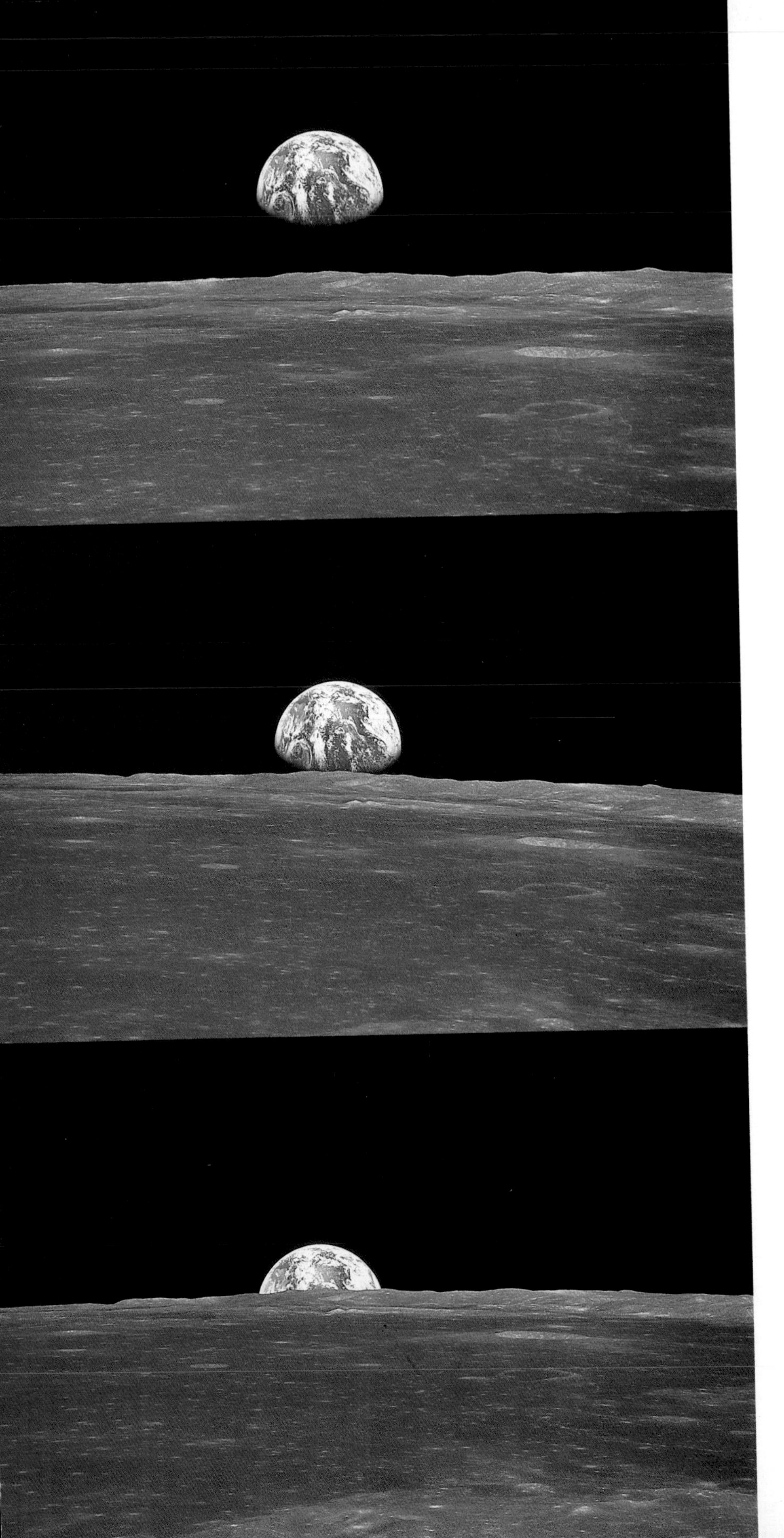

PAGES 2–3: The shuttle orbiter *Challenger*, being prepared for its maiden flight, is carried through the coastal fog from the Vehicle Assembly Building to Pad 39A.

PAGES 4–5: *Columbia*, the first of the orbiters, stands poised for launch on the night before its maiden flight, April 12, 1981.

PAGE 6: Astronaut John Young, flying at an altitude of about 20,000 feet in a shuttle training aircraft, photographs *Columbia*'s fifth launch as the rocket engines trail a long plume of exhaust.

PAGE 8: On Apollo 11, the first lunar landing mission, astronaut Mike Collins photographs the earth setting below the horizon of the moon.

PAGE 9: The orbiter *Challenger* executes this self-portrait after an experimental satellite has been deployed from its cargo bay on the seventh shuttle mission. A camera mounted on the satellite is operated by remote control, and the satellite is later retrieved by the orbiter's manipulator arm.

PAGE 10: The interaction of residual molecules with the orbiter's surface eerily lights *Challenger*'s OMS pods and tail on the eighth shuttle mission.

PAGE 11: On *Columbia*'s fifth flight, the author photographs the split-second burst of light that accompanies an ignition of the OMS engines.

PAGE 14: A heat-dissipating radiator mounted on the orbiter's cargo-bay door reflects the smooth curve of the earth below.

All photographs, except where otherwise noted, are provided courtesy of the National Aeronautics and Space Administration and belong to the public domain. Additional photo credits appear on page 220.

Edited by Leslie Stoker.

Distributed in the United States by Workman Publishing, 1 West 39 Street New York, New York 10018
First published in Great Britain by Orbis Publishing Limited, London 1984

Printed and bound in Japan.

ISBN: 0-85613-735-9

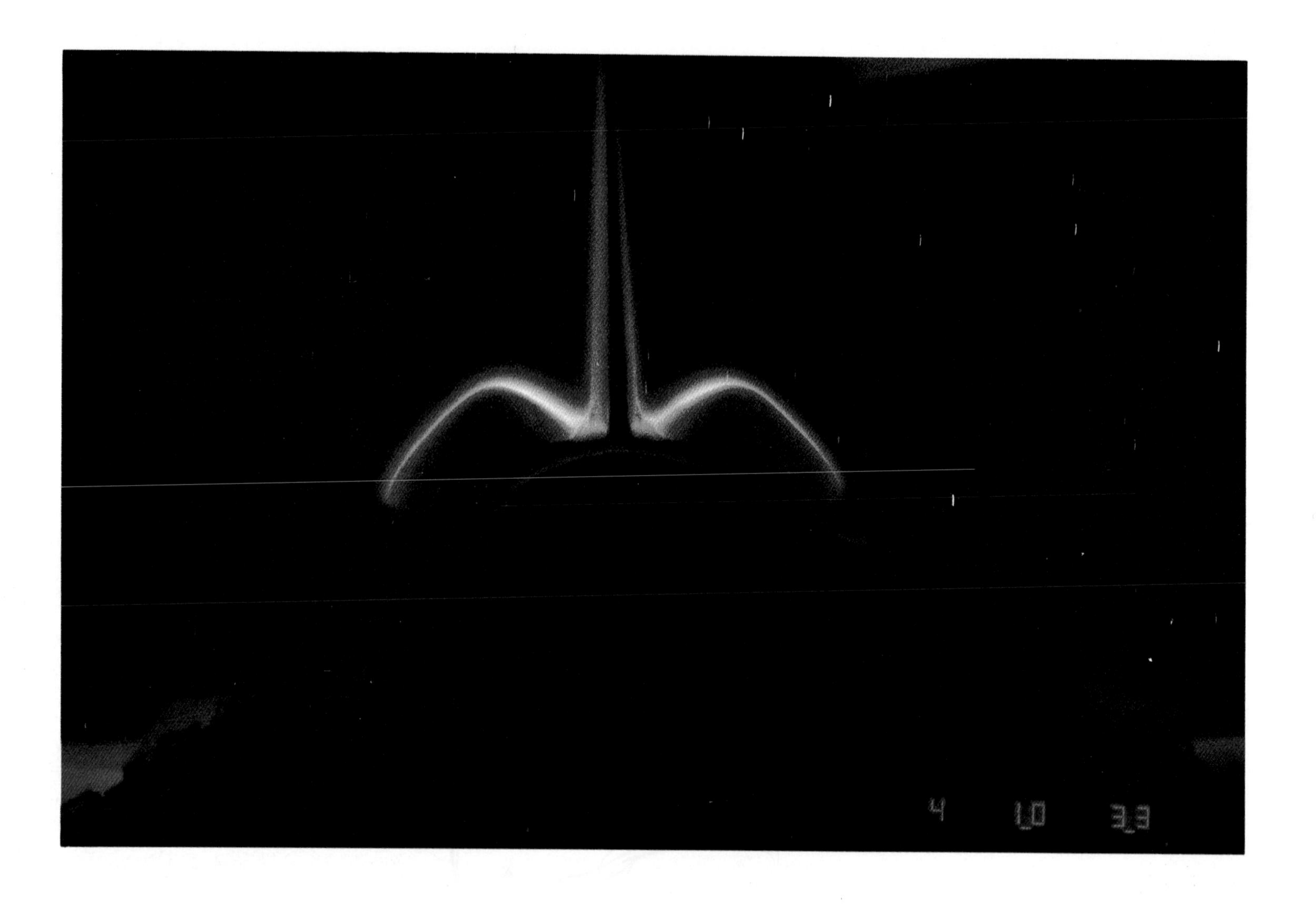

To be able to rise from the earth;
to be able, from a station in outer space,
to see the relationship of the planet earth to other planets;
to be able to contemplate the billions of factors
in precise and beautiful combination that make human existence possible;
to be able to dwell on an encounter of the human brain and spirit with the universe—
all this enlarges the human horizon. . . .

NORMAN COUSINS, 1973

PREFACE

I did not grow up wanting to be an astronaut. In those years there were no astronauts in Indiana, nor anywhere else for that matter. But there were real adventurers to read about—explorers, cowboys, race-car drivers—and I often longed to follow in the footsteps of Daniel Boone, Roy Rogers, or Sam Hanks. When I first crossed Sugar Creek in a homemade, square-transomed boat, I dreamed of the day I would be an explorer. But as I grew older and graduated to canoes, venturing downstream past the rock-strewn cliffs of Turkey Run and on toward the Wabash River, I began to suspect I was too timid to become a real explorer. Later, when I realized I was too small and not nearly musical enough to be a cowboy, and much too cautious to be a race-car driver, I decided to become a teacher—and perhaps someday I will be one.

In the meantime, extraordinary events happened. As I was studying Kepler's equations of orbital motion, they were dramatically demonstrated by the first orbits of *Sputnik*. As I was finishing my study of physics at Yale, John Glenn orbited the earth. Suddenly human imagination had made it possible to travel outside the earth's atmosphere at speeds *beyond* human imagination. Even more remarkably, scientists as well as pilots were selected to make these journeys, and I became part of the astronaut corps. I have since orbited the earth eighty-one times, traveling a distance greater than the entire length of the Wabash River.

This book is about the space experience, an experience made possible only by the dedication of thousands of Americans—technicians, engineers, and scientists—who lovingly create the machines that now circle the earth and cross the solar system. These spaceships, incorporating aspects of every science and applying every technology of our age, are more than computer-guided craft of cold composites and metal. They are monuments to human innovation. Aboard such a ship, in every corner, in every nook and passageway, we sense the skills of those who built it, and their minds and spirit travel with us.

It occurs to me that if Kepler were to read this book, he would not be surprised to learn that people in these spaceships orbit the earth exactly as his equations so elegantly predicted. But he would surely be awed to see the images of life aboard them and the views from them that we have captured with the modern invention called the camera. Most of the photographs printed here were taken with ordinary hand-held cameras, and they accurately represent what the human eye sees from space.

The photography experts at the Johnson Space Center in Houston, Texas, are unfailingly helpful in teaching astronauts to take photographs, and then in making their photographs available to the public after each mission is completed. In particular, I want to thank Dick Underwood, John Holland, Bill Robbins, Mike Gentry, Fernando Ramos, and Jeff Bremer.

I have greatly enjoyed looking at these photographs again and talking about space travel with Russell Martin, a friend, an author on the American West, and a novice in orbital mechanics. He has withstood my spacelog with astonishing patience, pointing out that at times astronauts tend to resemble cowboys, sitting in ejection seats instead of saddles, wearing helmets instead of wide-brimmed hats, and using transportation methods with millions of horsepower instead of one. I am indebted to Russell; to designer Hans Teensma, lover of space images and a storehouse of space statistics; to Leslie Stoker, our patient and persuasive editor, who is as careful with grammar as astronauts are with gravity; and to Andy Stewart, our enthusiastic publisher, who brought this unlikely team together and sustained us with an uncanny appreciation of the magnitude of space flight. They have made this book possible. Its art and elegance are theirs; any lapses in technical accuracy are mine.

Finally, I thank my wife, Bonnie, who has been supportive for many years, including those when space journeys were not as commonplace as they are today, and who encouraged me, having persisted this long, to write about my journeys beyond the earth. I am deeply indebted to her and to our children, David and Elizabeth, who, although bored by astronauts, want to become explorers, cowboys, and race-car drivers, and to do a host of truly exciting things when they grow up.

Joe Allen
Nassau Bay, Texas

CONTENTS

TO RISE FROM EARTH

An Introduction

The history of human travel into the sky and the space beyond is as brief and recent as it is bright with spectacle. People have dreamed of rising from the solid earth into the unknown for as long as they have inhabited the planet, yet less than two hundred years passed between the day in December 1783 when two Frenchmen first rose above a field outside Paris in a linen and paper balloon decorated with gilt fleurs-de-lis, and the day in December 1968 when three American astronauts first orbited the mysterious and long-enticing moon, a world away from our own. Fewer than eighty years elapsed between the first controlled flight of a winged aircraft over the sand dunes at Kitty Hawk, North Carolina, in 1903 and the first landing of a winged spaceship on a flat lake bed in California's Mojave Desert in 1981. And just fifty years separated the first backyard launch of Robert Goddard's small, hand-built liquid-fuel rocket in 1926 and the dramatic rocket launchings, in the 1970s, of a series of robotic spacecraft that landed on Mars, orbited Venus, and flew tantalizingly close to gigantic Jupiter and then on to Saturn, planets hundreds of millions of miles away from the earth.

The past two centuries hold the whole story of our ventures and investigations into the sky. And our challenges and accomplishments in space span little more than two decades. Twenty-five years ago we did not know whether we could construct a spacecraft that could survive the fiery friction of re-entry into the atmosphere. We did not know whether humans could endure the physical effects of weightlessness or the psychological effects of separation from the earth. We could not be *absolutely* certain that the moon was not made of green cheese, or that the half of the moon that is perpetually hidden from the earth was not inhabited by a civilization as old as ours. The possibilities of landing astronauts on the moon and safely returning them to the earth—or of sending a spacecraft to Mars to land on the cold red planet and dig in its soil for signs of life—seemed like distant and naive dreams.

PRECEDING OVERLEAF: In this photograph by the author, the orbiter *Columbia* approaches one of the sixteen sunrises it experiences daily while in orbit. The upper band of horizon is a false image created by the camera lens.

OPPOSITE: The shuttle orbiter *Challenger*, photographed by a satellite that it briefly deployed in July 1983, is brilliantly lit by the sun against the utter blackness of space.

Apollo 16 astronauts John Young and Charlie Duke prepare for their exploration of the moon in a lunar rover trainer at China Lake in the California desert. (James L. Long Associates)

But they were not. Within a decade, two American astronauts had walked on the surface of the moon, and ten others followed in turn. Within another decade, two spacecraft had safely landed on Mars and were conducting a sophisticated array of investigations there, and were transmitting wonderful photographs of that planet's spare, rocky surface and its thin and ruddy skies.

Today, even as the launches of astronauts on missions into earth orbit have become very nearly routine, and even as the Voyager 2 spacecraft races through the black void of the solar system toward a 1986 rendezvous with the planet Uranus, we are still in the first fundamental stages of entering space. We have only recently begun to reuse our spaceships, launching them like rockets and returning them to the earth like airplanes. We are only in the planning stages of constructing a permanent outpost in space. And we have just begun to shift the focus of space flight from exploration to operation—moving from experimental missions, designed to test how to travel safely to space and back, to operational missions, designed to take advantage of the unique properties of the vast realm beyond the earth.

An astronaut practices maneuvers to be used on the surface of the moon on the "rock pile" at the Kennedy Space Center in Florida. (James L. Long Associates)

The astronauts of the inaugural era of space travel were explorers—pioneers who ventured briefly out to the fringe of a limitless frontier. The astronauts who enter space today are their logical successors—men and women who go to space to go to work. They are pilots, scientists, and technicians who reflect our growing knowledge of and familiarity with space, workers who are increasingly at home in the new frontier.

The exploration and settlement of remote and unknown regions are elemental human activities. We have always been wanderers, searchers, explorers because we cannot quiet our deep curiosities. Yet our ventures into space represent a form of exploration that is fundamentally new. For more than 3 billion years, all life on earth has been held fast to the planet. The force of gravity has been the most elemental aspect of the earth's evolution. Only with our entries into space have we been able to break that basic rule, to take the first gigantic step toward the exploration of worlds other than our own. The Apollo missions to the moon—the first journeys of people from this world to another—

PRECEDING OVERLEAF and RIGHT: An Apollo/Saturn 5 launch vehicle is en route from Cape Canaveral's Vehicle Assembly Building to the launch pad. (Both: James L. Long Associates)

were perhaps the most important events in human history. Yet in another sense, those manned explorations of the moon were probably so inevitable at *some* point in our history that they belong simply to the long cycle of human discovery, spurred by the same impulses as the search for a passage to the Far East that led to the discovery of a new continent, or the journeys of mountain men and trappers to the headwaters of the great western rivers.

Any type of travel is a form of exploration, one that informs us about the new and unknown, and—just as important—offers us a fresh perspective and new understanding of the place where we started. Our quarter-century of travel into space has awakened us to the fact that we inhabit a fragile and finite planet on which we depend totally. Before people around the world had seen pictures of our planet taken from earth orbit or the startling photographs of it, small and blue, rising above the moon's gray horizon, it was difficult to grasp the truth that our home is just a sphere in space, a planet that orbits a rather ordinary star located in an inconspicuous corner of a minor galaxy. This truth is still hard to accept, but because of our first exploratory ventures into space, our perceptions of the fragile earth and its relation to the entire cosmos have been expanded, and we have greatly increased our knowledge of the planet that is our home and of the universe of which we are a tiny part. "Space is the new ocean," President John F. Kennedy said when the Mercury program got under way in the early 1960s, "and this nation must sail upon it." Space today remains an ocean that beckons us outward into the unknown, but it inevitably turns our vision back toward our small planet on its shore.

By 1958, when the National Aeronautics and Space Administration (NASA) was formed as a sudden response to the launching of the first *Sputnik* satellite by the Soviet Union, the Air Force had already been testing a series of supersonic, winged rocket-planes for more than a decade. The *Sputnik* scare convinced the Air Force that the time was right to proceed with the design of a rocket-plane that could orbit the earth, re-enter the atmosphere, and land like an airplane. The craft would be called the X-20, and it would be launched into space by a Titan III booster, a rocket that was still a few years away from being ready to fly.

NASA officials, however, pressured by the Eisenhower administration, Congress, and the public, were anxious to initiate a manned space program and were unwilling to wait for the development of the X-20 or the Titan III. They chose instead to send the first astronauts into space in tiny ballistic capsules mounted atop already-developed military attack

A technician models the shuttle space suit now in use. The special undergarment contains heating and cooling tubes that maintain a comfortable temperature. The legs, upper torso, gloves, and helmet provide air pressure around the body, and a backpack contains complete life-support systems. Umbilical lines no longer connect the astronaut to the spacecraft. (Anthony Wolff)

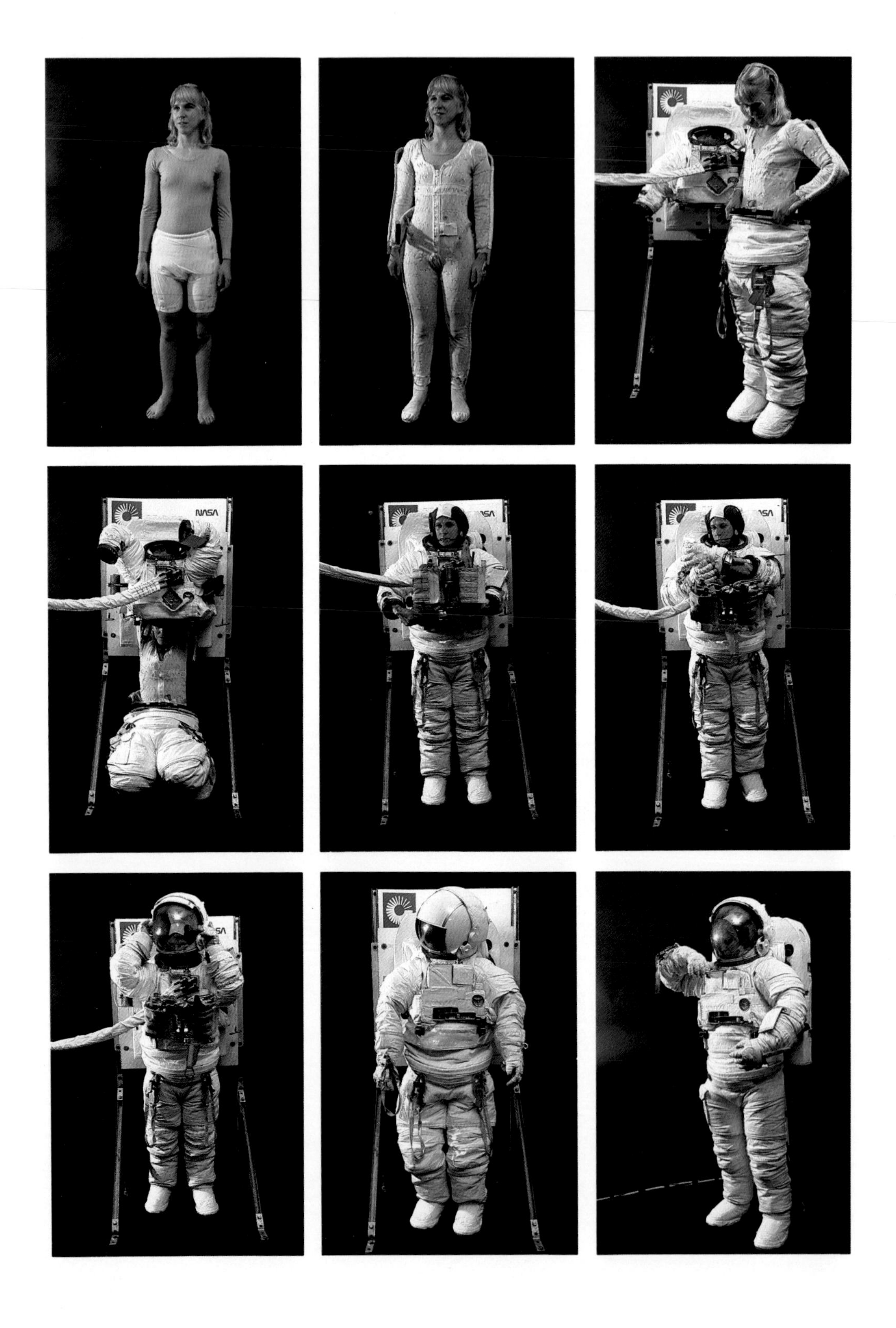

rockets. Once in space, the capsules would be slowed by retro-rockets, fall back through the atmosphere, and then be lowered by parachutes to an ocean splashdown.

OVERLEAF: A NASA 747 airliner ferries the orbiter *Challenger* to Florida for its maiden flight.

NASA's program to send an American into space was dubbed Project Mercury, and it quickly captured the attention and imagination of the entire country, so much so that the first seven Mercury astronauts became instant celebrities of heroic, perhaps mythic, proportions before they had even begun to train. By May 1961, the first of the seven had actually flown in space, and President Kennedy announced soon thereafter that the goal of the fledgling space program was to put an astronaut on the moon by the end of the decade.

Five more Mercury flights followed, then the ten two-man Gemini missions, then four preparatory Apollo flights before the historic July 1969 mission that sent astronauts Neil Armstrong, Edwin Aldrin, and Michael Collins into lunar orbit. Armstrong and Aldrin entered a lunar module, safely descended to the moon, and walked on its powdery surface. Six more Apollo missions to the moon followed this first landing, but NASA planners and project designers concurrently focused on the post-Apollo future of the American space program. Skylab, a three-mission program designed to test the physical effects of long durations in space on astronauts, was already being developed, and it seemed that the logical step after that was to create a reusable spaceship that would be able to ferry large crews and heavy payloads to space. This spaceship would serve as a simple space truck, eventually supplying a long-dreamed-of orbiting space station.

Early in 1972, the Nixon administration announced that the United States was initiating a new space program that would "take the astronomical costs out of astronautics" and would revolutionize space travel by making it routine, "turning the space frontier of the 1970s into familiar territory." Two months later, NASA officials decided that the new Space Transportation System would use a winged orbiter capable of at least a hundred missions into space. The orbiter would be launched by its own three rocket engines, which would be supplied with propellants from a huge external tank attached to the orbiter. Two partly reusable solid-fuel rocket boosters would help lift the 100-ton orbiter and the 830-ton tank from the launch pad and up through the thick, sluggish atmosphere. The orbiter would maneuver in space like the bell-shaped capsules that were its predecessors by periodically firing a combination of small, rocketlike thrusters. But it would then glide back into the atmosphere, using its wings and aerodynamic controls to land unassisted like an airplane.

This new "space shuttle" was, in many respects, an updated version of the moribund X-20 project that incorporated the enormous technological advances of the Mercury, Gemini,

Challenger

lenger
States
USA
NASA
905

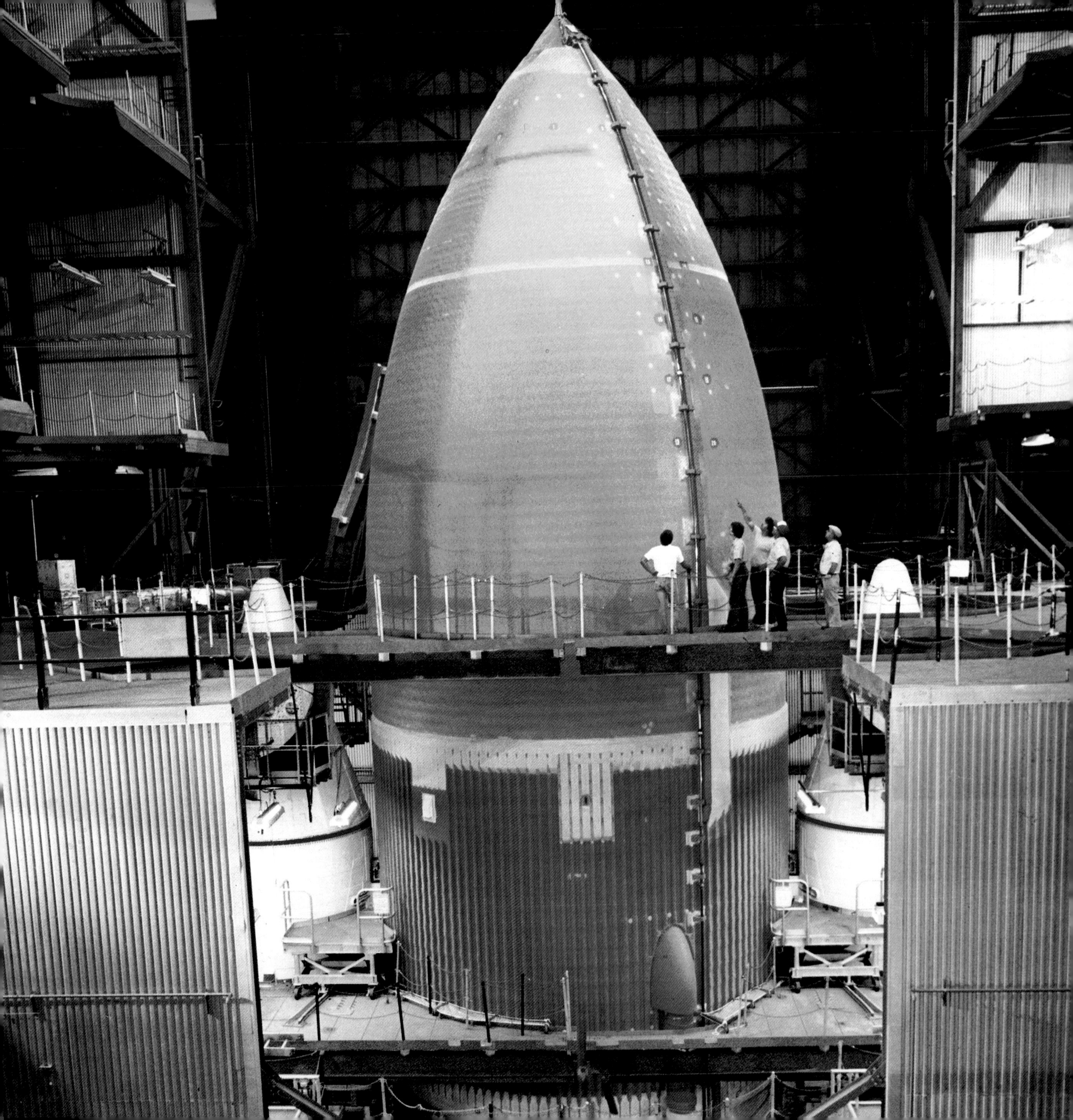

and Apollo programs. The shuttle would be the world's first true *aerospace* vehicle—equally at home in space and in the air—and it would represent the most complex single machine ever constructed. On April 12, 1981, the first test flight of the shuttle was successfully launched from Cape Canaveral, Florida. Two days later, the spaceship landed in California, setting smoothly onto a wide desert runway.

Inside the Vehicle Assembly Building, the two solid rocket boosters are mated to the fifteen-story-high external tank in preparation for the sixth shuttle flight.

What follows is an account in words and photographs of the human dimensions of modern space travel, its unique experiences and sensations. It is an examination of the machines that make the journeys into orbit and back possible, reliable, and safe—the thundering rockets, the sophisticated space guidance and maneuvering systems, the orbiter that can endure the fiery descent through the atmosphere, then land, and be launched again. And it is an examination of the personal experiences of the people who fly the missions beyond the earth—what they feel when the solid boosters ignite at launch and hurtle them suddenly into the sky; the pleasures and predicaments of weightlessness; the many jobs that each crew performs inside the orbiter and on the spectacular few-hour forays outside it; the stunning views of earth from orbit; and the dramatic hypersonic roller-coaster ride the crew experiences as the shuttle makes its meteoric descent into the atmosphere.

The astronauts who now enter space wear sport shirts and slacks during their days in orbit; they eat shrimp cocktail and barbecued beef and sleep in private bunks. The Spartan era of space travel has ended. The shuttle program has also expanded the roles of the astronauts. Astronauts who serve as pilots, mission specialists, and payload specialists perform a wider variety of intricate operations than did their predecessors, and they experience more directly the sublime environment of weightless space. And because of the ten large windows that surround the orbiter's flight deck, they can see and appreciate black space and the spinning earth more fully than was ever before possible.

The shuttle astronauts still prepare extensively for each mission; they relish their infrequent opportunities to fly; and they report that their travels to space have been their grandest journeys. Almost twenty-five years after the first astronauts were rocketed into the unknown, today's astronauts, now sojourners in familiar territory, still enter space with anticipation as high as the realm beyond the sky.

ENTERING SPACE

USA

The Launch into Orbit

A strange and marvelous flying machine is poised on its tail above the sand and scrub pines and sluggish deep-water lagoons of Florida's Cape Canaveral. The sun lights the Atlantic's far, flat horizon in the dim advent of dawn. Brown pelicans rise from their roosts and fly in formation above the surf; gulls arc over trees and dunes; swift ospreys lift up and begin to search for fish above the coastal sloughs. But the flying machine remains motionless. Just beyond the beach at Playalinda, it stands on Pad 39A like an enormous ruddy bullet flanked by two thin Roman candles, to which an intrepid white and black glider is attached.

PRECEDING OVERLEAF: In this early 1960s photograph, a Mercury/Atlas launch pad is lit on the night prior to a morning launch. (James L. Long Associates)

Every journey to space begins at the speed of one-half mile an hour. A crawler-transporter carries the shuttle launch assembly from Kennedy Space Center's Vehicle Assembly Building to the launch pad.

Strapped into seats in the nose of the delta-winged glider, five astronauts wait for their flight to commence. It is T minus 1 hour, 30 minutes. The commander, pilot, and mission specialists begin a series of preflight systems checks to make sure that this new breed of space vehicle is ready to begin another 2-million-mile journey. During the slow seconds and minutes between each check, the crew members look out the windows that surround the flight-deck consoles, watching waves break at the point where the ocean laps onto the land, watching shore birds climb and soar.

In the months that have led up to this morning's liftoff, the astronauts have flown hundreds of launches in flight simulators at the Johnson Space Center near Houston, Texas, transforming complex procedures into rote responses, preparing for every conceivable condition or failure, planning contingent reactions to myriad emergencies. In the final days before liftoff, the crew—its preparations virtually complete—has been sequestered in an effort to prevent last-minute exposure to colds and flu and the crushing attention of reporters and distant relatives. During the isolation, crew members have adjusted their circadian rhythms to the sleep-rest cycles planned for the mission, discussed final changes

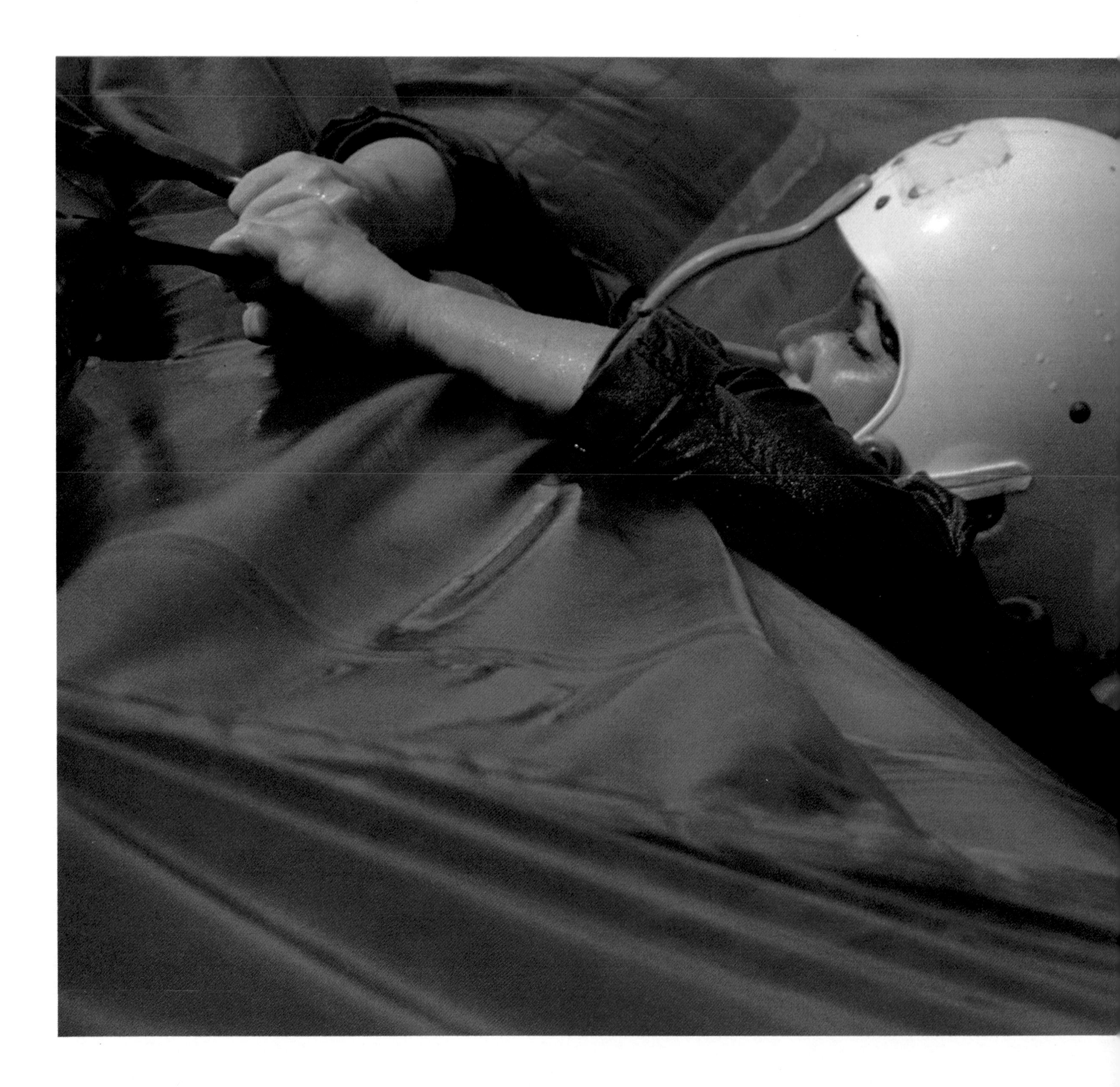

LEFT: Astronaut Judy Resnik in water-survival training near the Kennedy Space Center complex in Florida. (James L. Long Associates)

ABOVE: Rescue crews practice lowering an incapacitated astronaut from a mock-up of the orbiter's overhead escape hatch to the water. In the background, the orbiter *Columbia* stands ready for its fifth launch. (James L. Long Associates)

Astronauts Guy Bluford and Mike Smith arrive at Patrick Air Force Base near the Kennedy Space Center in a T-38 training aircraft. (James L. Long Associates)

OPPOSITE: At the mate-demate facility at Dryden Flight Research Center at Edwards Air Force Base in California, a shuttle orbiter is prepared to be lifted so that the 747 to which it will be mated, visible in the background, can be rolled into place. (Roger Kallet)

OVERLEAF: Astronauts Deke Slayton, Tom Stafford, and Vance Brand, accompanied by suit technicians, approach a crew-transport van during training for the 1975 Apollo-Soyuz Test Project, a successful rendezvous and docking of an American and a Soviet spacecraft. (James L. Long Associates)

and additions to the flight plan, and then flown in small, supersonic T-38 training aircraft from their homes in Texas to the east coast of Florida. Housed in spare, no-frills quarters at the Kennedy Space Center near Cocoa Beach, the crew relaxes, re-examines the checklists, notes, and procedures in the massive flight-data file, and receives formal briefings on global weather conditions and the preflight status of the launch vehicle.

Then in the final hours before the flight, isolated from all but their immediate families and the NASA officials and fellow astronauts who are designated "primary contacts," the crew members simply wait. Years of general preparations are behind them, and the long, repetitive training period in which they have learned mission-specific responses is now over. The reward for their labors lies just ahead.

The commander privately studies the flight plan; one crew member makes phone calls to friends and family; another reads. Then, in the thick and sultry darkness of the Florida night, the crew drives to the pad to view the vehicle in its dramatic launch configuration. The white orbiter, roughly the size of a DC-9 commercial jet, is lit by gigantic searchlights. It is

NASA
NO SMOKING

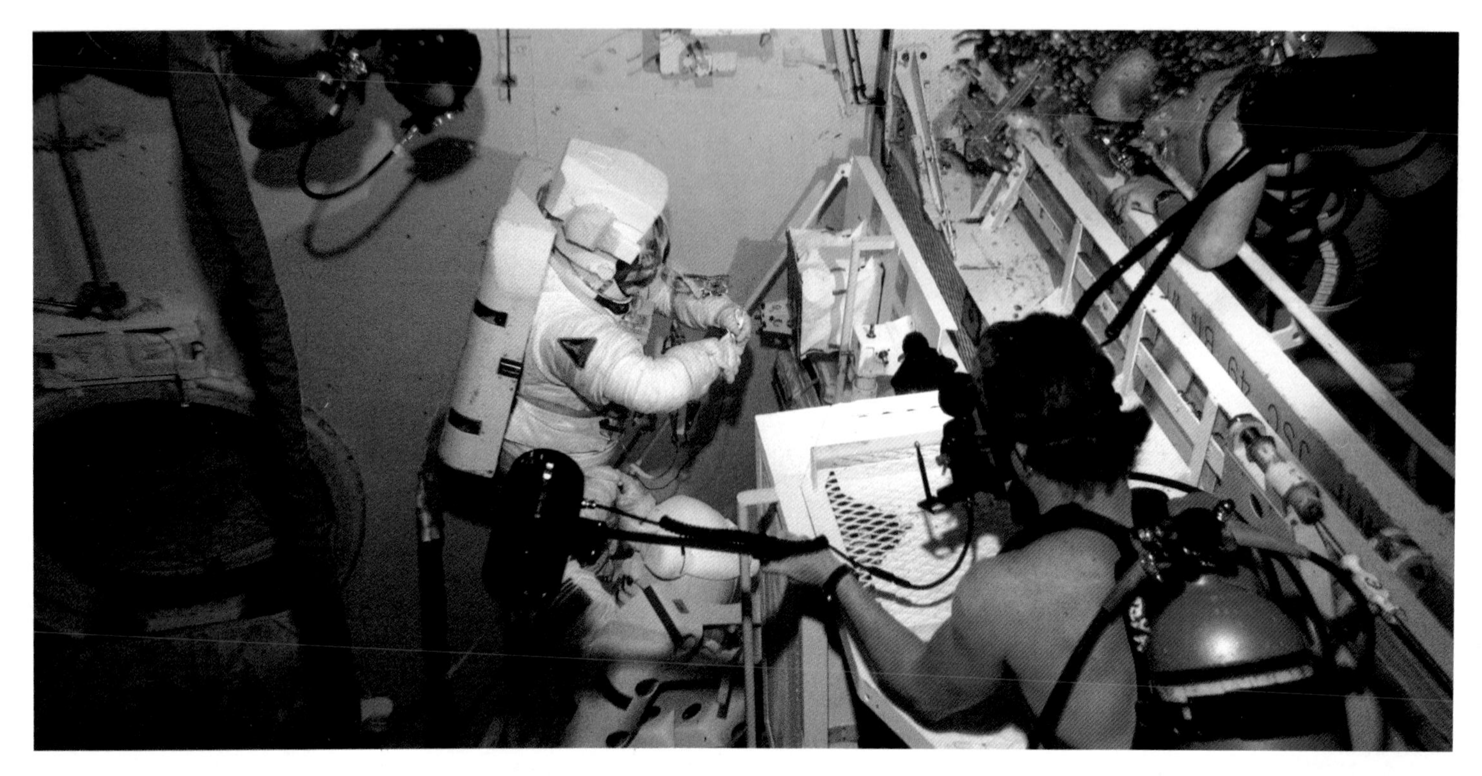

ABOVE: Shuttle astronauts Bill Lenoir and the author train underwater for a spacewalk at the Weightless Environment Training Facility at the Johnson Space Center.

RIGHT: The author is suited for underwater training. The "Snoopy helmet" he wears holds his communications ear-pieces and microphone. (James L. Long Associates)

OPPOSITE: A shuttle astronaut, experiencing temporary weightlessness in a KC-135 training aircraft, practices inserting a foot into a foot restraint, a device enabling a person to "stand" in weightlessness.

SPACE SHUTTLE

bolted to a fifteen-story-high liquid-fuel tank that is flanked by two enormously powerful solid-fuel rockets. The astronauts relish their reflective look at the flying machine—the only unhurried view they will have of the poised launch vehicle in its entirety. Then they return to crew quarters to sleep—or to try.

The crew is awakened about five hours before the scheduled launch. Each crew member dresses in high-topped boots and a fire-resistant one-piece flight suit adorned with a large name tag, a crew patch designed specifically for this mission, and an American flag. The commander, the only space veteran on this crew, wears special astronaut wings above his name tag. The era of launching astronauts in bulky, oxygenated Buck Rogers suits has ended; this crew wears clothing much like a military flight crew's. No longer are astronauts suited up like brave bullfighters preparing to enter the ring; no longer must they ceremonially consume steak and eggs in front of a television camera, as they did in the early space-age re-enactments of the Last Supper. Today, following a more normal breakfast, they spend time stuffing pens, pocketknives, sunglasses, calculators, flashlights, food sticks, and flying gloves into dozens of flight-suit pockets. Their wallets and personal belongings are collected for safekeeping, along with preissued visas ready for rapid delivery to countries where the crew could land if serious problems arise and the mission has to be ended prematurely. The astronauts then leave their quarters, smiling and waving dutifully at the reporters and camera crews who still record the ritual of each launch, and board a van that carries them through several security checks to the pad.

Technicians wait at the base of Pad 39A to escort the crew to an elevator in the service tower—a tall, gray steel support structure much like those that were called gantries during the Mercury, Gemini, and Apollo programs in the early years of space travel. The slow elevator stops at Level 195, where a long, steel-grated catwalk leads to the white room that abuts the open hatch of the orbiter. Technicians in this sterile room, dressed in lint-free coveralls, caps, and booties, help each crew member into a vest-like escape harness—a harness that can be grasped quickly, in the event of a fire or sudden explosion, to pull a disabled astronaut out of the orbiter and away from the service tower.

The impatient crew members have little to do in the white room but wait until the technicians begin to load each of them into the orbiter like last-minute payloads—commander first, pilot next, then mission and payload specialists. Each is methodically and carefully assisted into an assigned seat—helmet on, straps tight and buckles secure, legs and feet in place. The seats look remarkably like padded kitchen chairs, tipped over backward so

Columbia is poised on Pad 39A prior to the launch of STS-3 (the NASA abbreviation for Space Transportation System, mission number 3).

At sunset on the evening before the launch of STS-5, *Columbia* stands ready. (James L. Long Associates)

that the crew members lie flat on their backs, the best position in which to bear the forces of acceleration of the imminent launch.

As the commander is boarded, the rest of the crew cannot help but notice that the vehicle has at last come to tenuous life—groaning, rumbling, smoking—as if it were only seconds away from liftoff. This is certainly no flight simulator, no training mock-up; it is a new and somewhat startling machine, now almost alive. The cryogenic hydrogen and oxygen, so cold they are in a liquid state, began to be pumped into the huge external tank at about the time the astronauts were awakened. Now the volatile propellants boil and fume, contracting the aluminum skin of the tank and venting gases through a cap in its nose. The vehicle is finally a ready rocket that commands enormous potential energy, but for now that energy remains in a captive chemical form.

When the last crew member is in position and secure, the technicians make their final cabin checks and unplug their headsets from the ship's intercom system. They can no longer speak to the tightly helmeted crew, but their pats on the shoulder, handshakes, and thumbs-up signs convey their good wishes. They then leave the orbiter and secure the hatch behind them. While the technicians complete their work, the crew is already beginning its final preflight checks—checking communications with Launch Control at the Cape and with Mission Control in Houston, checking cabin pressurization, and monitoring dozens of systems on cathode-ray tubes, console lights, and gauges. The final hour of the countdown moves at a pace that is quick but not frantic, yet there is a kind of relentlessness to the countdown that demands that each prelaunch event or check proceed almost precisely as scheduled. This launch has been planned to occur during the three-minute period from 7:19 A.M. to 7:22 A.M.—a "launch window" that is determined by the position of the earth relative to the sun, the destinations of the satellites loaded on board, and the desired lighting conditions at the intended landing site eight days from now. None of these factors will tolerate human tardiness.

At T minus 51 minutes, the Inertial Measurement Unit, a sophisticated system of computer-assisted gyroscopes, is aligned—meaning that it is informed of the vehicle's precise position relative to the rest of the solar system. The computers respond, noting that the vehicle is positioned at sea level, on lines of longitude and latitude that correspond to Pad 39A on Cape Canaveral, and is spinning eastward, due to the earth's rotation, at about 1,000 miles an hour. At T minus 30 minutes, the technicians in the white room and the remaining ground crew leave the service tower and move to a spot 3 miles away where they can safely

view the launch. Only the crew, strapped in the poised vehicle, remains at the pad, now watched by numerous television systems and sound-activated cameras.

At T minus 20 minutes, the final mission data are entered into the orbiter's on-board computers, there is a last launch-abort signal check, and then there is a ten-minute hold to allow all systems to synchronize before the final count. Following this hold, and only after a final weather clearance has been issued and all systems appear to be functioning properly, the countdown is resumed at T minus 9 minutes. The white room and catwalk swing away from the orbiter 7 minutes before the launch, and the crew prepares to start the auxiliary power units—hydrazine-fueled power systems that drive hydraulic pumps. The pumps move all the aerodynamic control surfaces of the orbiter and enable the engines' nozzles to gimbal, or swivel, in response to commands from the orbiter's guidance computers, steering a course into space by minutely adjusting the engines' angles of thrust, the engine nozzles in effect balancing the enormous vertical rocket like a yardstick on the tip of a finger.

The pilot starts the auxiliary power unit at T minus 5 minutes, and he and the commander initiate a quick series of hydraulic-pressure checks. When they have verified that the hydraulic systems are functioning normally, the main engines are gimbaled into a preprogrammed series of movements as a final test of the system, and the motion of the nozzles increases the trembling of the poised machine. The orbiter switches from its ground power system to its on-board electrical system. Vents in the external tank close, and the liquid oxygen and hydrogen, suspended in separate chambers, begin to pressurize.

The launch director instructs the crew to lower the visors on their helmets and to turn audio volume controls up high to allow communications to continue amid the tumultuous noise of liftoff. The crew feels the vehicle shudder again as the main engines gimbal into final launch position, their flying machine now anxious to break away. But only when the countdown reaches T minus 30 seconds and continues—as management of the launch switches effortlessly from the Launch Control computers to the on-board computers, severing the final electrical link to earth—is the crew certain that the launch is about to commence.

At T minus 8 seconds, thousands of gallons of water, stored in a tower that flanks the pad, are suddenly released in a Niagara of spray into a depression at the pad's base to dampen the sound energy from the engines, which will soon ignite and cause reverberations that could severely damage the orbiter's wings and tail and fragile payload. Three seconds later, signals go out from the on-board computers, opening valves that permit the supercold

STS-5 rises above an enormous cloud of steam and rocket exhaust about 10 seconds after liftoff on November 11, 1982. (James L. Long Associates)

OVERLEAF: Birds at Cape Canaveral are startled by the early-morning launch of STS-7.

liquid hydrogen and its oxygen oxidizer to travel through fat umbilical hoses to the orbiter's three main engines, where the liquids are turned to gases, compressed, mixed, and ignited to produce a sudden thrust of 1.12 million pounds in the combined engines—enough power at that moment to supply the state of New York.

As the main engines ignite, the crew feels a gigantic growl of fire rise from far below them and a sudden sense of motion as the vehicle lurches nearly 2 feet laterally in the direction of the external tank from the thrust of the flaming engines. At T minus 2 seconds, the orbiter's computers automatically check to confirm that the main engines have reached proper pressurization and thrust. Then, as the nose of the orbiter returns to a vertical position, igniters near the noses of the two solid-fuel rocket boosters send long bursts of flame into the core of the highly combustible powdered-aluminum propellant. Within half a second, the flame has spread throughout the 1.1 million pounds of solid fuel in each booster, forcing white-hot, pressurized exhaust gases through the engine nozzles and spewing them out of the exit cones with a combined thrust of 5.2 million pounds—more than four times the thrust of the three liquid-fuel engines and almost enough power to light up all the

Mercury *Friendship 7*, piloted by astronaut John Glenn, is rocketed toward the first orbit of the earth by a manned American spacecraft on February 20, 1962. (James L. Long Associates)

Faith 7 and its Atlas rocket rise above the original launch complex at Cape Canaveral on May 15, 1962. Astronaut Gordon Cooper piloted this final Mercury mission. (James L. Long Associates)

Atlantic states. The boosters, the largest solid-fuel rockets ever constructed, ignite with an incredible booming and crackling roar. Once in flame, they cannot be turned off or throttled back as liquid-fuel rockets can. As the solids ignite, explosive charges release the eight bolts that hold the whole assembly onto the pad, and the vehicle at last lifts off, trailing two 600-foot-long fountains of sun-bright exhaust from the solid boosters, and three pale blue, almost invisible, streams of exhaust from the orbiter's main engines.

Ground crews and spectators watch the launch vehicle rise from this inferno of smoke and steam, searing light, and earth-quivering noise. The astronauts on board see little but the service tower, which suddenly seems to drop away. But they *feel* the thunder of their lurch into the air with their whole beings—the numbing noise, the torrent of power exploding below them, and the exquisite, crackling acceleration.

The journey from earth to orbit, from the teeming Florida marshlands to the still vacuum of space, takes 8 minutes, 50 seconds. It has been planned and anticipated by the crew for at least six months and is the product of more than a decade of research and development of the Space Transportation System program. It resembles the

A fountain of sun-bright exhaust trails *Columbia* during the first 2 minutes of the fifth shuttle mission. (James L. Long Associates)

long exploratory voyages of Magellan, Drake, and Lewis and Clark in earlier centuries, yet it is a journey that reaches its destination in only a few mind-stunning minutes.

For the crew, the journey begins with a rush of noise and surge of motion. An instant after the tower disappears from view, the astronauts feel the launch vehicle roll into a high arc over the Atlantic, taking advantage of the earth's own thousand-mile-an-hour eastward motion, the orbiter suspended beneath the external tank, the crew in a heads-down position. The vehicle accelerates to a force of 3 Gs (three times the force of earth's gravity) within seconds after launch, and it quickly becomes impossible to distinguish any direction but forward. As the rockets lunge, the crew members are pushed back hard against their seats; their bodies feel very heavy, and they can move their arms only with deliberation. The vehicle vibrates wildly, and in spite of the crew's protective helmets, the cabin is incredibly noisy. Within a span of 30 seconds, the blue sky outside the orbiter's windows has become a deep and solid black.

As the vehicle approaches a speed of about 700 miles an hour, 50 seconds into the flight, the orbiter's main engines are throttled back to 65 percent of their potential thrust to prevent overstressing the orbiter's windshield area, wings, and large vertical tail by the growing pressure of the rushing air. As the vehicle climbs and the atmosphere gets thinner, this aerodynamic pressure decreases and the engines are again throttled up to full thrust. Only a minute later, the shuttle is traveling faster than Mach 4 (four times the speed of sound), hurtling nearly 3,000 miles an hour, 28 miles above the earth.

The crew feels a sudden deceleration as the solid-fuel rocket boosters burn themselves out 2 minutes into the flight. Then a startling flash of light signals that the spent rockets have exploded away from the tank. Small rocket motors push the two rockets cases away from the tank and orbiter, and the cases coast upward behind the speeding spacecraft before finally slowing, arcing, and falling back toward the earth. Parachutes inflate while the cases are about 4 miles above the Atlantic, slowing their descent to the ocean about 175 miles east of Cape Canaveral. There, tugs wait to tow them back to the space center to be reconditioned, refueled, and ultimately refired.

Free from the crackling tumult of the solid-fuel rockets, the crew's ride suddenly seems much quieter and surprisingly smoother, as if the main engines were simply huge, humming electric motors. At 6 minutes, 30 seconds, into the flight, at a speed of Mach 15 (about 11,000 miles an hour) and at an altitude of 80 miles, the engines gimbal the orbiter and its tank into a shallow dive in preparation for engine cutoff and tank separation. Two minutes later, the engines are again throttled back to 65 percent of thrust, and as the orbiter reaches 17,000 miles an hour and continues to accelerate, the commander prepares for the shutdown of the main engines. He calls "MECO" ("main engine cutoff") as the computers issue the command to extinguish the throttled engines, and the crew members feel themselves moving forward in their straps from the sudden absence of thrust. Sixteen seconds later, explosive charges automatically blow apart the bolts that connect the orbiter to the enormous, empty tank that looms above it. Five reaction-control motors in the nose and tail of the orbiter ignite, carefully easing it away from the tank, much like someone in a rowboat would back away from a thrashing whale. None of the crew members can see the tank in its position against the orbiter's underbelly, so the jettison procedure is entirely automated, executed precisely to prevent a potentially disastrous bump or collision.

When the orbiter and tank have safely separated, residual oxygen in the tank is vented to control its tumble into the atmosphere, where the heat of re-entry causes the tank to break up into fiery chunks. Its debris falls into a 600-mile-long "footprint" in the Indian Ocean.

Following separation from the tank, at an altitude of about 70 miles, the vehicle remains fractionally too slow to continue into orbit. Its two orbital maneuvering system engines, mounted on either side of the aft fuselage near the main engine cluster, fire a 2½-minute burst of hydrazine fuel oxidized by nitrogen tetroxide. The crew feels the orbiter accelerate again, increasing its speed by about 130 miles an hour, just enough for it to reach its necessary orbital velocity of 17,590 miles an hour. The first ignition of the orbital maneuvering system engines—known as the "OMS-1 burn" in the acronymic jargon of the space workers—lifts the vehicle into an elliptical orbit that reaches an altitude of roughly 170 miles by the time the orbiter circles halfway around the earth, a point that is less than 45 minutes away. As the orbiter, now a bona fide spacecraft, reaches the apogee of its elliptical orbit high over the South Pacific, the maneuvering engines ignite for a second, somewhat shorter time—the OMS-2 burn. This ignition shifts the craft into a steady, repetitive, circular orbit around the earth. The flying machine now coasts in silent space.

At dawn on April 12, 1981, twenty years to the day after Russian cosmonaut Yuri Gagarin became the first person to enter space, astronauts John Young and Robert Crippen rocketed away from the watery landscape of Cape Canaveral on the first test flight of the space shuttle. The fledgling Space Transportation System program was initiating the era of reusable spacecraft, an era in which the fiery lunges of manned

FAR LEFT: STS-7 begins its "roll" maneuver seconds after liftoff. The dark trail of exhaust streams from the solid rocket booster engines, and pale exhaust pours from the three main engines aboard the orbiter *Challenger*. (James L. Long Associates)

LEFT: The power required to lift the shuttle-launch assembly beyond the atmosphere is nearly equivalent to the electrical power needed to light the Eastern Seaboard for those brief minutes. (James L. Long Associates)

rockets would become very nearly routine. But in the twenty-year history of human journeys into space, there had never been a first launch that involved so much unproven machinery and rocketry—or so much redundant, fail-safe computer technology.

No unmanned test flights of the shuttle launch vehicle had ever been conducted. The shuttle orbiter was the largest and heaviest single vehicle ever placed into orbit in a single launch, the first winged craft ever to reach speeds even approaching 17,000 miles an hour. Almost entirely automated at critical phases of its launch—each of its four redundant computers able to perform 325,000 operations a second—the orbiter was undoubtedly the most complex machine ever built. In its launch configuration, it was the first manned rocket system to deviate radically from a slender, cylindrical design. Its boosters were the first solid-fuel rockets to lift a human cargo.

A coastal lagoon mirrors the launch of the tenth shuttle mission in February 1984. (Otis Imboden)

Young and Crippen wore pressurized safety suits on that first flight, and they were strapped into ejection seats that appeared almost as lethal as the potential disasters they were designed to allow the astronauts to escape. But as the orbiter's main engines flared to life that morning, as the solid-fuel rockets exploded in a cascade of searing exhaust, and as the launch proceeded in perfect cadence, the astronauts marveled at how much smoother the ride was than they had expected it to be. They noted that the solid-fuel boosters had performed even better than expected, thrusting the vehicle 10,000 feet higher than anticipated before they burned out. It was Crippen's first space flight, and when the engines shut down at MECO, he was amazed by the view of the earth below, splendid visual proof that the launch had been a success.

Nearly four years after the maiden launch of the orbiter *Columbia,* the Space Transportation System has become operational; launches now follow one another with quick and reassuring regularity. The ejection seats are gone; the pressure suits are no longer worn. Weather requirements have been relaxed, permitting launches through cloud layers and at night. The once-empty cargo bays of the three operational orbiters are now full on each flight, and crews of as many as six people are commonly lifted into space.

Yet for both veteran and novice astronauts, launches will never become routine. They will always remain so technologically complex, so cooperatively challenging, and so personally exhilarating, that each will be one of the richest events of a lifetime. For each crew member, there will always be the anxious and interminable waiting, the stunning moment of ignition, the thrill of acceleration, and the silent surprise of sloe-black space. There will always be the marvel of seeing the earth from orbit, the wonder of having escaped its bounds.

THE BRIGHT COCOON

PRECEDING OVERLEAF: Skylab, a large orbiting experimental laboratory, was inhabited by three crews for a total of nearly six months in 1973 and 1974.

OPPOSITE: Astronaut Walt Cunningham peers wearily through the window of the command module near the end of the eleven-day Apollo 7 mission—the first to test the machinery that would ultimately go to the moon.

The atmosphere that encircles the earth is as thin as an onion skin, yet its swirling film of gases supports and sustains the planet's life. But outside the atmosphere, beyond the earth's moist and enveloping blanket, life can be maintained only in fragile, artificial environments. Each spacecraft that abandons the shelter of earth and enters the vacuum, the emptiness, called space must be a kind of cocoon in which conditions on earth are imitated, in which temperature is controlled and food, water, and air containing a breathable amount of oxygen are supplied. Every spacecraft must be a vessel built in the image of the earth but one that is also able to survive the rigors of a realm vastly different from earth.

Temperatures in orbit around the earth can range from as high as 250° F, when a spacecraft is exposed to the direct rays of the sun, to less than –150° F in absolute blackness, when the sun is obscured by the earth. There is no air in space, no atmospheric pressure, no wind or weather or sound. Yet a spacecraft does have to contend with potentially damaging solar flares, cosmic radiation, micrometeoroids, and abrupt and dramatic changes in temperature. Since the first Mercury capsules were launched a quarter-century ago, spacecraft have been designed with external skins that can withstand those conditions that are literally otherworldly and that can seal in their interior earth-like environments. The shuttle orbiters now in operation employ an array of advanced technologies to provide crew members with short-term space homes that closely approximate the earth's natural environment and its contemporary living conditions. A single orbiter—an aluminum-alloy airframe covered with aluminum panels that are surfaced with ceramic and fiber insulation—houses forty-nine engines, twenty-three antennas, and five computers. Two independent and redundant systems produce pressurized air comparable to the earth's air at sea level, and heat, cool, filter, and circulate it. Three independent and redundant power-generating systems combine cryogenically stored oxygen and hydrogen in fuel cells to produce dependable electricity and abundant water as a convenient, and quite drinkable, by-product.

But the one condition of life on earth that cannot be duplicated in the orbiter, or in any

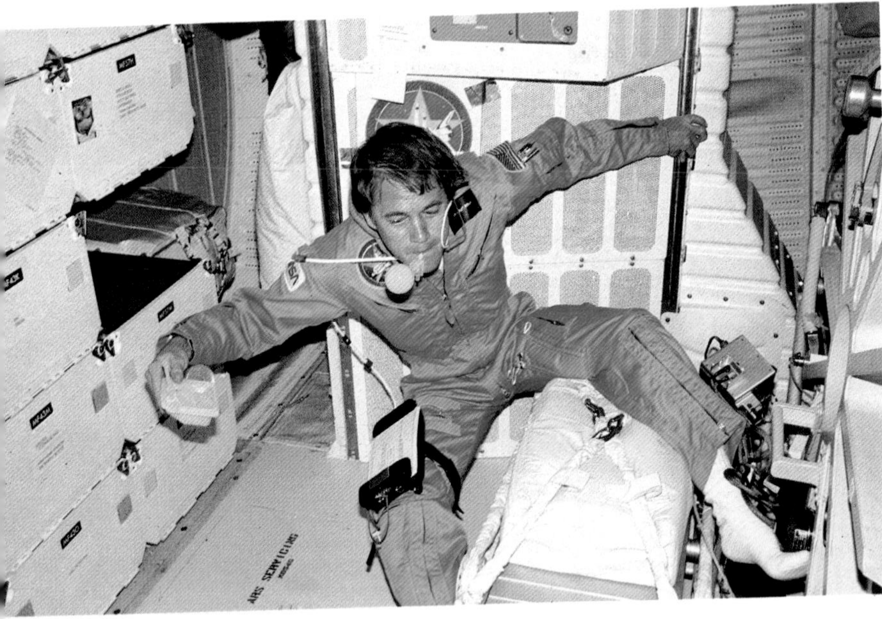

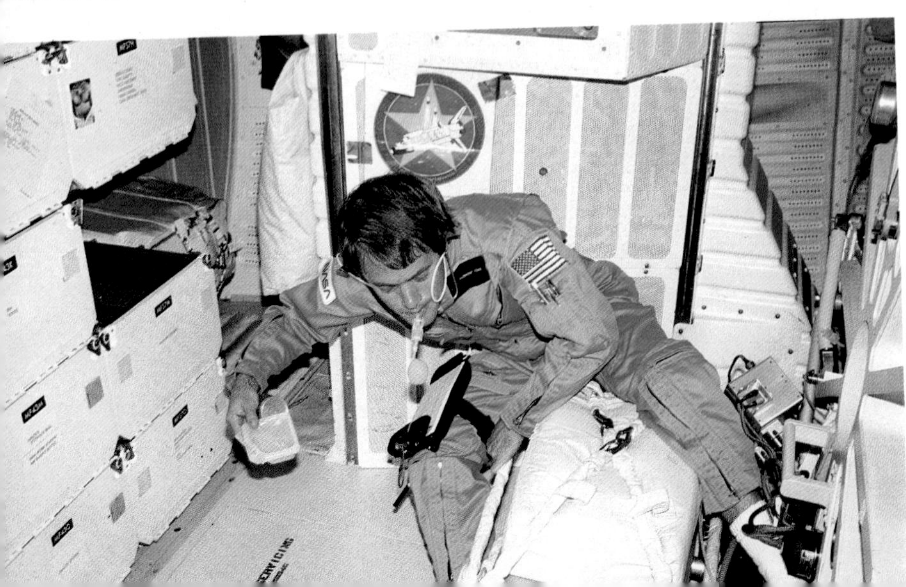

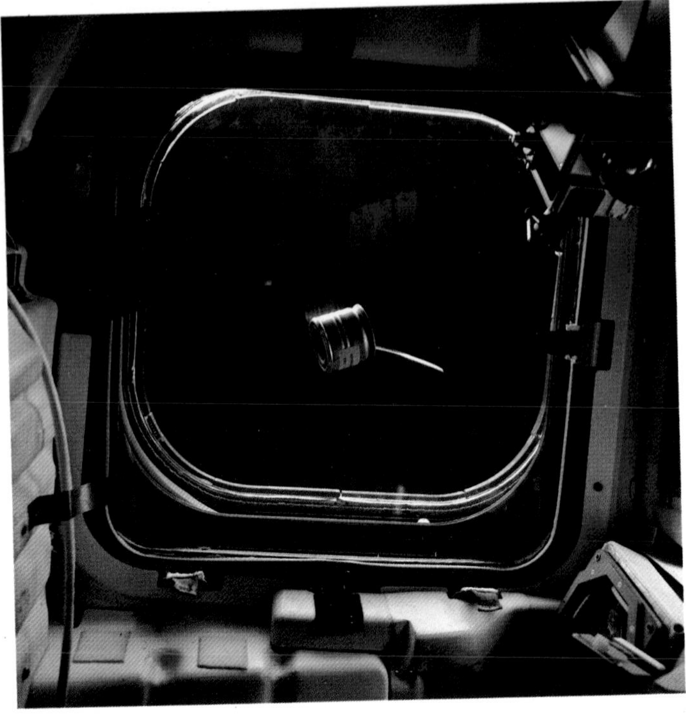

ABOVE: A spoon and a container of diced pears float in front of the orbiter's overhead window.

LEFT: The author, photographed by Bob Overmyer, chases down a floating sphere of orange juice and sucks it up with a straw.

spacecraft, is the effect of earth's gravitational pull—the inescapable force that gives weight to every object, and keeps it anchored to the planet. A substitute for earth's gravity *could* be provided to space travelers in orbit around earth or, for that matter, on their way to a distant planet, by causing their spaceship to spin, as did the large space station in Arthur Clarke's *2001*. Objects and crew members inside a spinning craft would be held to the outer walls of the vehicle by centrifugal force, the same force that holds water securely in a bucket swung over one's head. In such a spacecraft, astronauts would stand with their feet against the outer surface, now a legitimate floor, with their heads pointing in toward the axis of spin. But because a major reason for being in space is to utilize the weightlessness found there, as yet there seems to be no reason to go to the additional effort to engineer and build a spinning space habitat just to make its occupants feel more at home.

Gravity does not somehow mysteriously disappear in orbit, however. It is the gravitational tug of the earth, in fact, that holds spacecraft, satellites, and the moon itself in precise and eternally predictable orbits. The phenomenon of weightlessness called "microgravity" by physicists and "zero gravity" by astronauts, results from a balance between the earth's gravitational pull and the inertia of a spacecraft—that property (first recognized by Newton in 1666) that, without the countering gravity, would carry the speeding craft forever farther from the earth, always in a straight line. At an orbital velocity of just over 17,500 miles an hour, a spacecraft has enough speed to prevent gravity from pulling it back to the surface of the earth. Yet the earth's incessant tug does prohibit the craft from racing out of the planet's grasp into the vast emptiness of the solar system.

The orbit of the shuttle spacecraft around the earth is a very literal kind of constant falling, producing for its crew the same effect of weightlessness as one can experience momentarily in an acrobatic airplane dive or in the stomach-stunning plunge of a roller-coaster. But the weightlessness of earth orbit is perpetual. There is no pulling out of the airplane dive, no body-compressing shock at the bottom of the roller-coaster track. Zero gravity is floating without having to pay the consequences—a strange and sublime experience that is at once bizarre and immensely enjoyable, a relaxed, slow-motion state in which all of the earth-bound rules have been broken. And zero gravity renders the ordinary events of daily life in orbit endlessly interesting, transforming them into complicated and comic acts and producing dozens of pleasurable and perplexing situations. Without weightlessness, living in space would be much like living aboard a ship or submarine. But the inescapable and intriguing reality of weightlessness makes living in space like nothing else imaginable.

Astronaut Bruce McCandless photographs the orbiter *Challenger* during the first experimental flight of the manned maneuvering unit, a rocket chair that allows astronauts to fly free of their home in space.

During its 4-million-year evolution, the human body has been shaped, sized, and engineered by the gravitational conditions of life on earth. Skeletal structure and musculature provide support against the stresses imposed by the body's weight; the size of the heart and the intricacies of the circulatory system reflect the need to pump blood against the force of gravity out to the arms and up to the head. In orbit, however, the physical demands placed on the space travelers are markedly different. The pilot who weighs 190 pounds on earth suddenly weighs nothing at all. The mission specialist who is 5 feet 4 inches tall on earth *grows* an inch or more because her vertebrae are no longer compressed by her body's weight. Fluids are redistributed in the weightless state, causing each face to look flushed and a little puffy; the muscles in the crew members' legs—which are now virtually useless and unused—quickly begin to atrophy; appetites are suppressed, and heart rates slow significantly.

In the first hours of weightlessness, almost half the astronauts who enter space experience a strange and temporary malaise that is now called space-adaptation syndrome. This condition was at first presumed to be a weightless form of motion sickness, but physician-astronauts are now beginning to guess that this flulike lethargy, occasionally accompanied by vomiting but not by nausea, is grounded in an eye-brain response to the startling absence of *up* or *down*—an adverse reaction to the loss of earthly equilibrium. Traditional motion sickness, suffered by air and sea travelers, is caused largely by a disturbance, a kind of sloshing around, of fluids in the semicircular canals—the tiny motion-sensing organs of the inner ear. Space-adaptation syndrome, however, seems to be related only indirectly to semicircular canal disturbance. And the mystery has been deepened by the fact that several astronauts have reported that their condition worsened when they viewed, from the cabin of the orbiter, the earth's horizon in sudden, unexpected, and disorienting positions. In any case, the symptoms of the malaise invariably disappear within a day or so.

Although astronauts have always privately and cautiously admitted to varying sorts of space malaise, only in recent years, with the number of space voyagers rapidly increasing, has the phenomenon received careful attention. The first series of American spacecraft—like their Soviet counterparts—were so small and confining that their crews were literally squeezed, prodded, and pressed into their seats, a process technicians called insertion. Aloft in orbit, the pioneer astronauts scarcely had space enough to move their legs or arms, and any sort of floating about their capsules was plainly impossible. For the crews of the Mercury, Gemini, and even, to some degree, the Apollo missions, weightlessness and its atten-

The author photographs Vance Brand and Bob Overmyer, commander and pilot respectively of STS-5, looking out the aft windows of *Columbia*'s flight deck to confirm that the cargo-bay doors are closing properly.

OPPOSITE: In this photograph by the author, astronaut Bill Lenoir eats dinner in the aft flight deck of *Columbia*.

dant pleasures and predicaments were experienced only peripherally. Those first astronauts experienced genuine weightlessness, but because they were strapped into seats, they did not feel free flotation, and their heads were virtually always upright in relation to the configuration of their capsules' interiors. Perhaps for those reasons, the malaise seemed to be less of a nuisance to them than to their successors in space.

It was not until the three Skylab missions in 1973 and 1974 that astronauts in orbit were finally able to explore the phenomenon of floating in open and unencumbered environments. Skylab, an experimental space station fashioned from a converted Saturn rocket booster and assorted Apollo spare parts, had as much habitable living area as a three-bedroom house. Designed specifically for long-term missions, Skylab was enormous. Long-

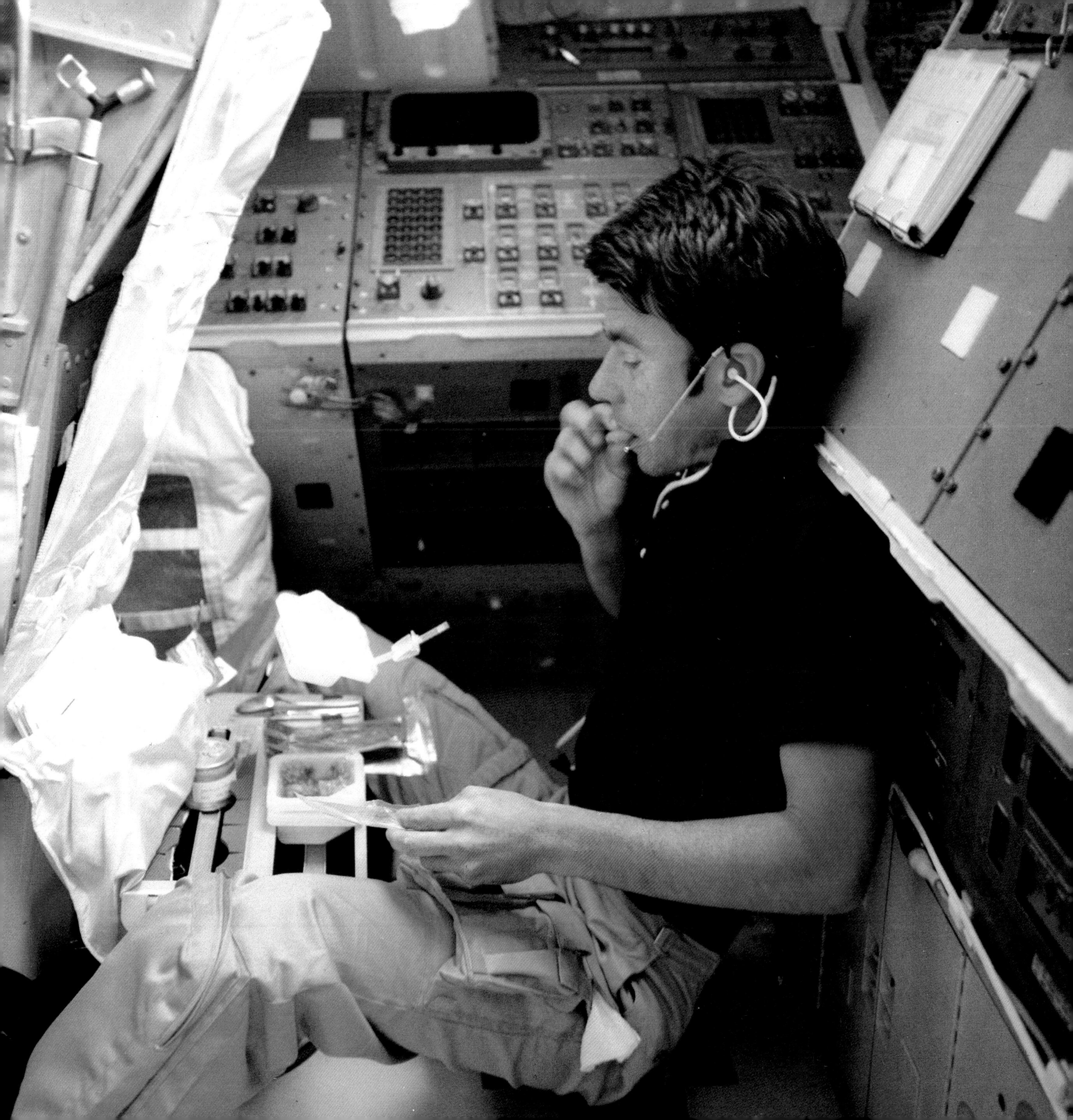

PAGES 71, 72, and 73: The orbiting laboratory Skylab was occupied by three different crews during 1973 and 1974. A second rectangular, electricity-generating solar panel would have flanked the long cylindrical living quarters had it not been torn away in the launch.

er than the height of a twelve-story building, Skylab contained almost 12,000 cubic feet of living space, compared with a mere 210 cubic feet inside a cramped Apollo command module or the tin-can-like confines of a Mercury capsule, whose bell-shaped interior measured less than 4 feet by 5 feet.

The three manned Skylab missions lacked most of the grand adventure that characterized the Apollo moon landings that preceded them. Yet in many ways the Skylab missions were more taxing physically and psychologically than previous space flights had been. The Skylab program's principal focus was the investigation of how well the human body could withstand long stretches in space. Although there was certainly no *a priori* reason to believe the human body could adapt to long periods of living and working without gravity, the preliminary indications from Skylab were that, with proper precautions, it could do so surprisingly well.

When Skylab 2, the first manned Skylab mission, was launched on May 25, 1973, the longest period American astronauts had spent in space to date was the fourteen-day flight of Gemini 7 in 1965. Flight surgeons who examined astronauts Frank Borman and James Lovell after that flight discovered that both astronauts were weakened because of loss of muscle tissue; loss of calcium had begun to deteriorate their bones; both had lost substantial amounts of body fluids, including blood; and the blood itself was low in red cells, electrolytes, and hormones. Bits of information about Soviet flights increased concerns about the dangerous effects of long-term weightlessness: Three cosmonauts who had been in orbit for eighteen days in 1970 were so debilitated that they had to be carried from their spacecraft when they landed.

However, the crew members of Skylab 2—Pete Conrad, Paul Weitz, and Joseph Kerwin, the first astronauts who actually *lived* unencumbered in space—returned from their twenty-eight-day mission in much better condition than their Russian counterparts had, probably because the Skylab crew had orbited inside a very large craft and had been able, and were in fact encouraged, to exercise regularly. Although their sense of balance had been affected by their month in space—they had trouble walking without wobbling on the deck of the U.S.S. *Ticonderoga* after splashdown—they nonetheless could ambulate. Curiously, Conrad walked better and said he felt better than he had immediately following his ten-day Apollo 12 mission.

During the course of his four-week sojourn in Skylab, Conrad had often exercised on a stationary bicyclelike device called an ergometer. Weitz had used it occasionally; Kerwin, a

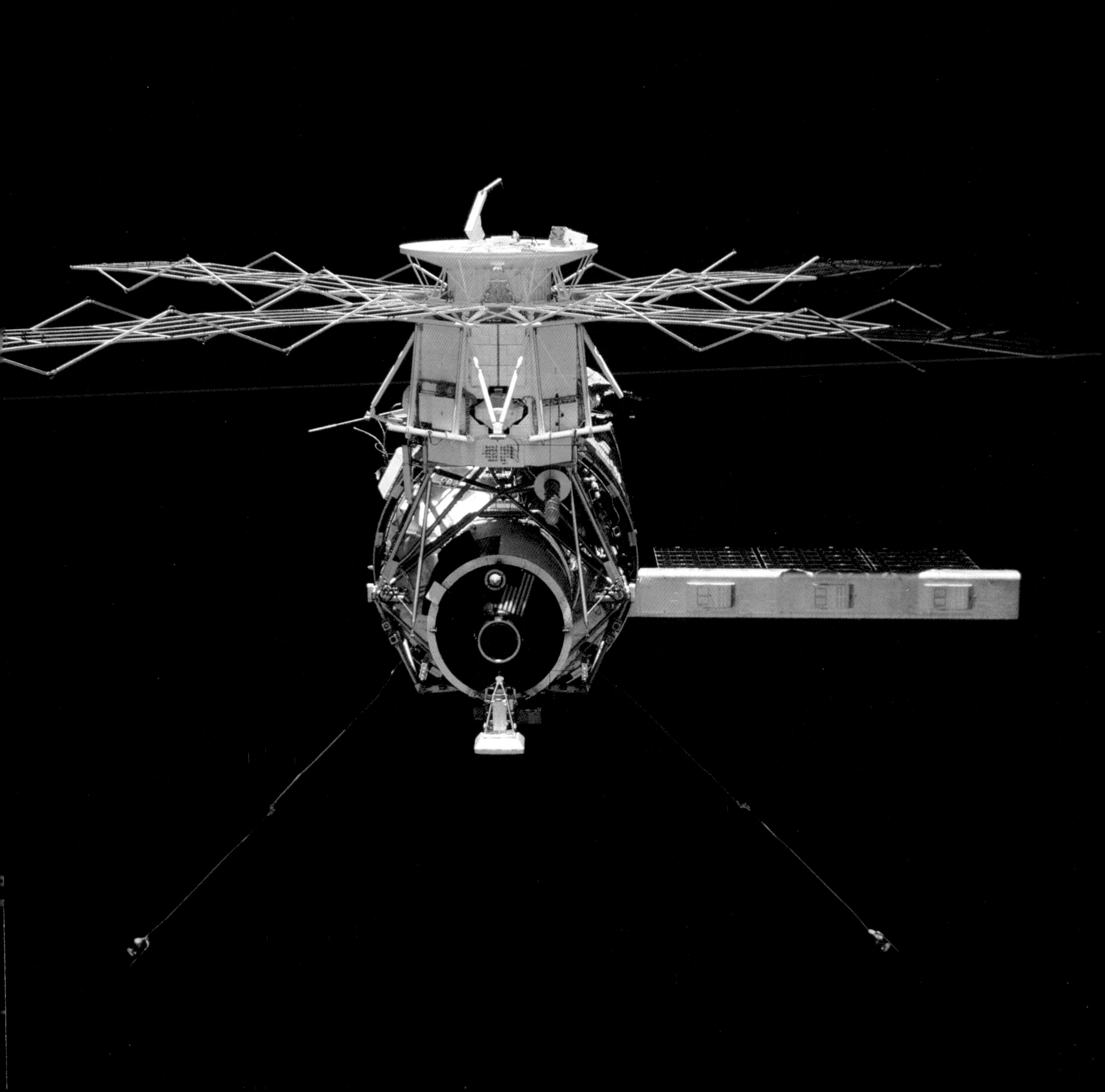

physician, seldom. Since each crewman's postflight condition seemed to correlate with the amount of time he had spent exercising, flight surgeons demanded that the crews on subsequent Skylab missions exercise regularly and vigorously—a requirement that seemed to lead to dramatically better conditioning. The crew of Skylab 3, who orbited the earth for fifty-nine days, twice as long as Skylab 2, was estimated by flight surgeons to be in about 20 percent better condition than the first crew. The final Skylab crew, whose mission lasted just short of three months, returned in the best condition of all.

The medical evidence derived from that last Skylab mission suggested, rather tantalizingly, that although the body does change in a weightless environment—and in some potentially harmful ways—it does stabilize and ultimately adapt remarkably well to an environment without precedent in human evolution. It began to seem possible that astronauts could conceivably spend six months, a year, or even longer in space.

At the time of the last Skylab mission, a new program that was increasingly being called the "space shuttle" was already in advanced stages of planning and design. The shuttle orbiter itself was too small to offer spacious living to large crews in orbit, but its name implied that it was being designed merely to *shuttle* astronauts between earth and an outpost in orbit—an outpost that would be a kind of permanent address where people could really *live* in the weightless, worldless realm of space.

The bilevel cabin of today's shuttle orbiters—a flight deck that contains the cockpit and crew station, from which the vehicle is flown and its payloads managed, and a mid-deck that houses the craft's food-preparation, sleeping, waste-disposal, and storage areas—has a combined living volume of 790 cubic feet, fifteen times smaller than Skylab's. But aloft, at least, the orbiter's interior dimensions become truly three-dimensional. For instance, the 13-by-10-foot mid-deck, which appears cramped for a five-person crew prior to liftoff, when it is packed with stowage compartments, bunks, and launch seats, becomes relatively roomy in orbit, when its seats are stowed and its ceilings and walls are every bit as habitable as the floors.

The distinctions between ceilings and floors, between walls and ceilings, disappear amid the weightless drifting. Any direction is *up* and any surface is *down* only in relation to the eyes and mind of an individual. Yet most crew members floating inside the orbiter prefer to keep their bodies oriented toward the cabin's own up-down configuration, if only to make it easier to read dials and switches, or to pull out and rummage through storage drawers looking *down* at them rather than *up* into them.

But regardless of whether the mission specialist keeping house is upright or sideways in relation to the deck's ceilings and instrument panels, there is nowhere in the entire cabin to set an object *down.* The security of being able to put something on a shelf or table, knowing that it will stay there until one picks it up again, vanishes in the floating environment. Nothing has weight to hold it in place, and tables, shelves, and even floors are useless. A food tray will not stay on a table, nor will the pilot's boots sit politely on the floor. Every object that needs to stay in position has to be hooked, clamped, kept behind a compartment door, or held in place with Velcro. The inside of an orbiter in flight, in fact, is vividly etched with Velcro strips and loops of adhesive tape—patched-on anchor points that hold food containers, cameras, clipboards, and crew members themselves in stationary positions long enough to accomplish a given task.

Together with the perpetual drifting, the continual need to attach every movable object lends the most menial job an aura of great deliberateness. Tasks that are virtually reflexive on earth demand careful concentration in space. Otherwise a few ounces of juice suddenly begin to float about the cabin like a small orange planet; pens sail along the circulating air currents; and toothbrushes roam maddeningly out of reach.

Such mishaps are frequent in the first few days of every space flight. Watching inexperienced crews trying out their "space legs" is like watching a small skating rink crowded with adults who are on ice-skates for the first time. During the first days in space, the act of simply moving from *here* to *there* looks so easy, yet is so challenging. The veteran of zero gravity moves effortlessly and with total control, pushing off from one location and arriving at his destination across the flight deck, his body in proper position to insert his feet into Velcro toe loops and to grasp simultaneously the convenient handhold, all without missing a beat in his tight work schedule. In contrast, the rookies sail across the same path, usually too fast, trying to suppress the instinct to glide headfirst and with vague swimming motions. They stop by bumping the far wall in precisely the wrong position to reach either the toe loops or the handholds. In their attempts to recover before rebounding to the starting point, they twist around too rapidly, knocking loose cameras, film magazines, food packages, and checklists, all previously Velcroed to the wall and now careening about the cabin in different directions. This feeling of awkwardness and the associated frustration of trying to hurry in a world that demands that all actions be in slow motion gradually subside. By the third day, rookie and veteran are moving with equal skill.

Before people had orbited the earth, and before anyone had directly experienced weightlessness, it was widely presumed that eating in space would be a bizarre and difficult experi-

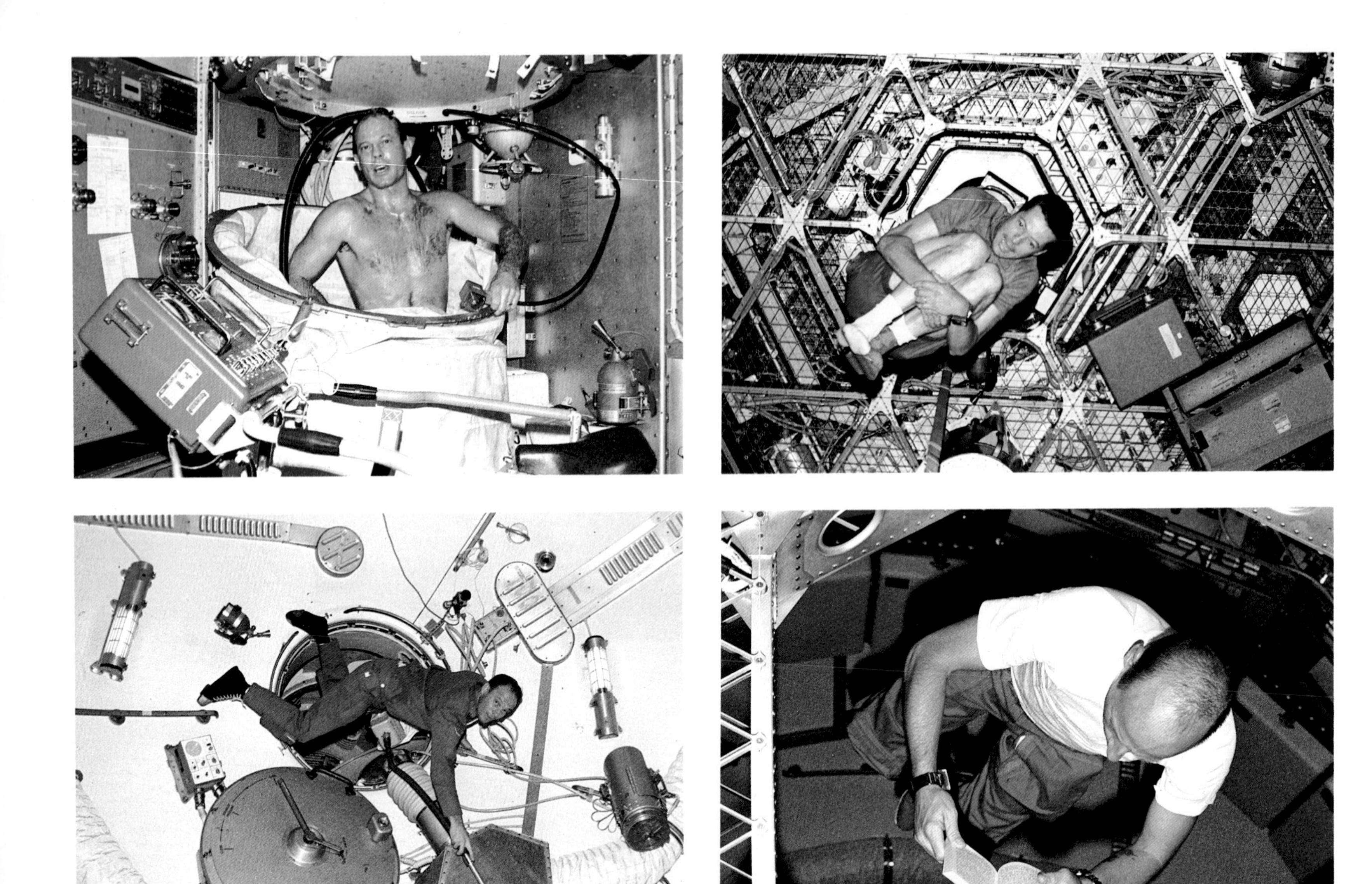

CLOCKWISE FROM TOP LEFT: Skylab astronaut Jack Lousma demonstrates the zero-gravity shower; Joe Kerwin tumbles effortlessly in Skylab's storage area; Pete Conrad floats while he reads; and Jack Lousma vacuums an air filter that is part of Skylab's air-conditioning system on the dome of the storage area.

ence. Bread, it was guessed, might break into thousands of pieces; vegetables would disintegrate; soups might explode out of their bowls. Meals on the first flights were restricted to items that could be pureéd and placed in squeeze tubes or compressed into chewable tablets. Eating was a rather distasteful endeavor that resembled ingesting baby food out of a toothpaste tube.

Experience in orbit quickly proved, however, that eating in space was much like eating on earth. No foods ever exploded or fell to pieces in space tests. Once opened, canned foods tended to stay inside their cans; puddings, crackers, and candy bars posed no special prob-

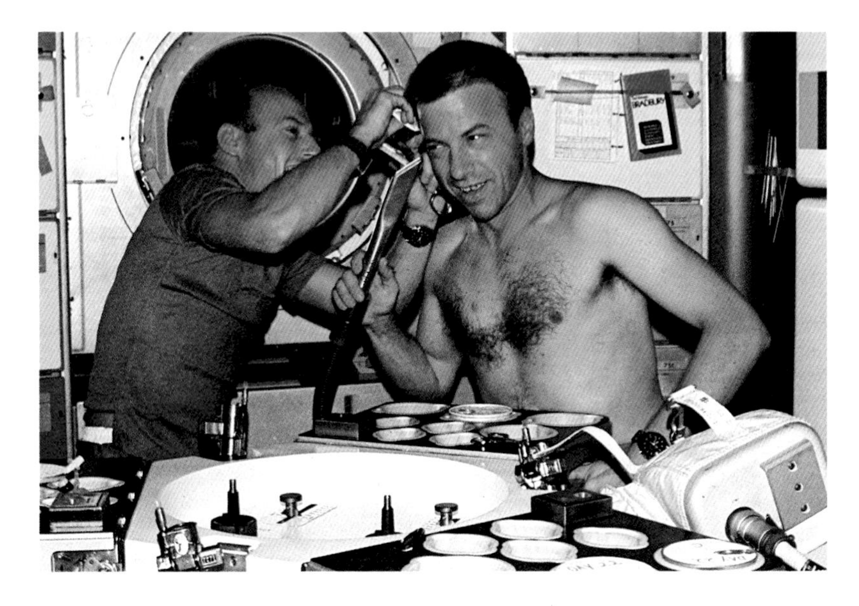

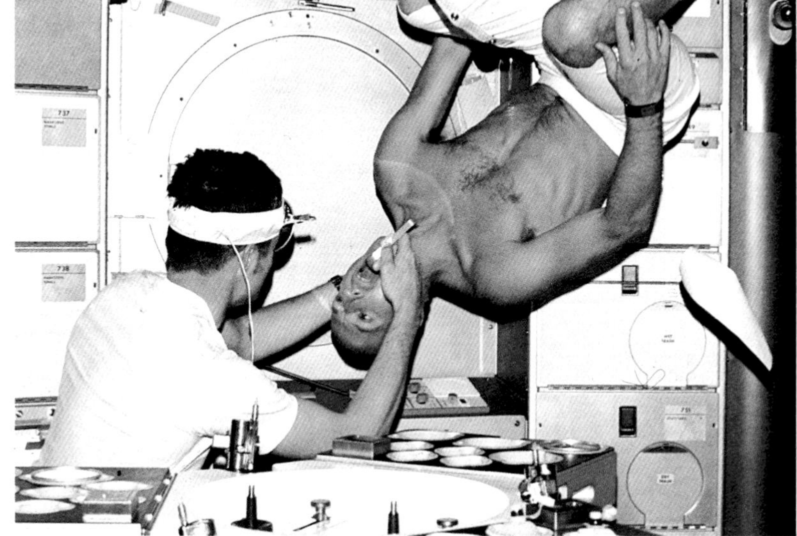

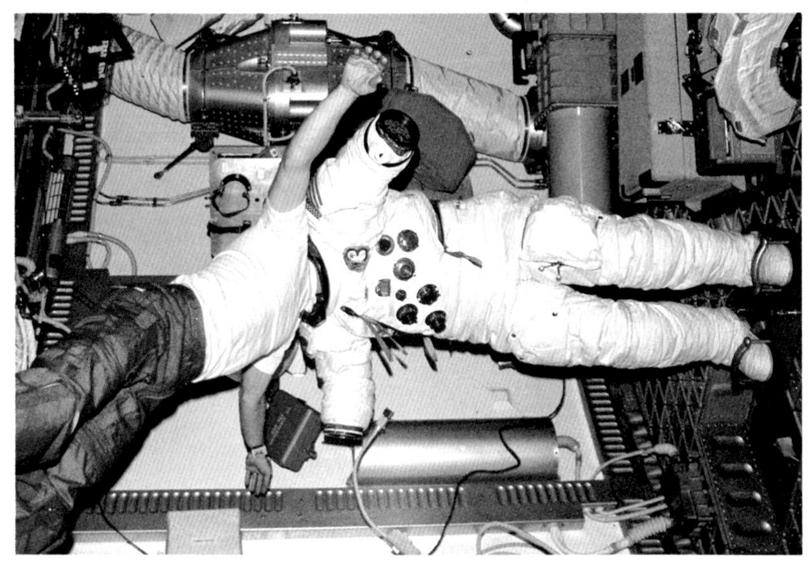

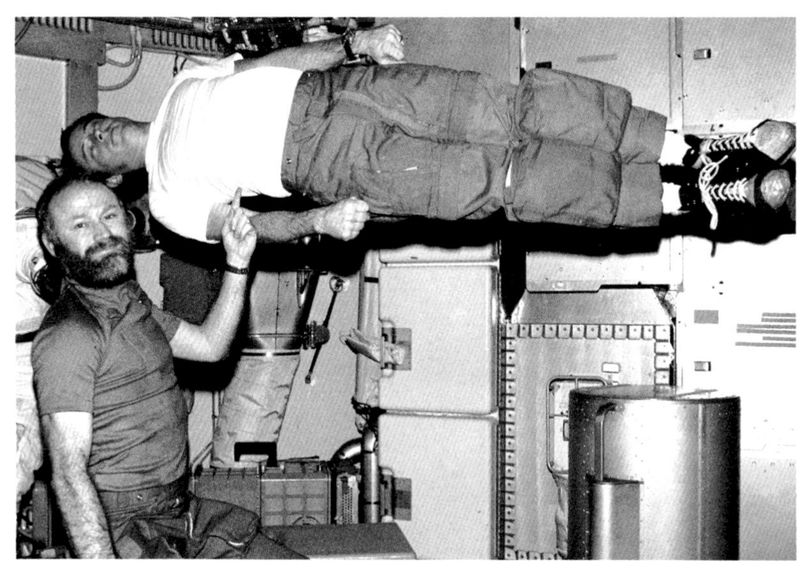

CLOCKWISE FROM TOP RIGHT: Dr. Joe Kerwin gives Pete Conrad a dental checkup; Gerry Carr demonstrates how easy it is to lift Ed Gibson, who doesn't weigh anything; Ed Gibson inspects a space suit; and Pete Conrad gives Paul Weitz a haircut.

lems. Today, each space-shuttle mission carries a pantry containing almost a hundred different kinds of food and twenty different beverages—from asparagus and apple juice to turkey tetrazzini and tropical punch. Except for plastic-wrapped slices of bread, virtually all food items are freeze-dried or dehydrated to save weight; the orbiter produces its own water, so it need not be carried from earth. Beverages are powdered or freeze-dried as well, so there is no beer or liquor on board. (Mission planners for the three Skylab flights did intend to include a stock of table wine for the crews' enjoyment, but the Women's Christian Temperance Union objected, and NASA, always mindful of its public image, scuttled the plan.)

LEFT: On board Skylab, "Arabella," a space spider, weaves a zero-gravity web. Although the spider initially seemed disoriented by weightlessness and its webs were erratic, it soon adapted and began to create normal web patterns.

ABOVE: A photograph taken from the dome of Skylab's storage area shows astronaut Gerry Carr reading a checklist, an empty space suit held in place by a foot restraint, and the passageway into the living quarters.

The shuttle orbiter's kitchen is a compact closetlike galley located on the mid-deck. It includes hot- and cold-water injectors, serving trays, a food-storage pantry, and a hot-plate-like oven. Food packages, of course, will not sit placidly on this warming unit, so it is nested into a small aluminum suitcase that is tethered to the end of an electrical cord and floats aimlessly through the mid-deck. To prepare a meal, a crew member simply injects an indicated amount of water into each plastic single-serving container, shakes its contents to mix the water and dried food, then places the container in the suitcase to heat.

Each serving tray holds containers, cans, and silverware in place—a means of keeping the meal manageable. Although none of the food has any weight, its own surface tension tends to keep it inside its open container; only a hard bump and the laws of physics will send chicken à la king careening across the cabin. Conventional utensils are used to lift food from containers to mouth—except that spoons do not have to be held right-side up. Since there is no "up," a spoon can hold its cargo at any angle, and a dropped spoon simply drifts, chocolate pudding still attached, until recovered by its rightful owner. The processes of chewing and swallowing are no different than they are on earth, and the space food itself, while unlikely to win any culinary awards, is perfectly palatable. Cleaning up after meals is a simple matter of depositing empty food containers in the trash compartment and wiping off serving trays and utensils so they can be used again.

Beverages pose a bit more of a problem. Liquid will not slide down the edge of a tilted glass or cup, so a straw is required to drink water, juice, coffee, or cocoa. Dozens of plastic drinking straws are stowed in the galley. Each straw features a special clip or closure that can pinch off the flow of liquid, which once set in motion would otherwise continue until the container was empty.

The very behavior of all liquids in zero gravity is so different from that of liquids on earth that the mere sight of water, grape drink, or orange juice out of its proper container borders on slapstick comedy. For example, strawberry drink squeezed from its container remains on the end of the straw attached to the container as a perfectly motionless round red globe. But if a curious crew member disturbs it slightly by shaking the straw or by blowing on it, the globe begins to shiver, tiny tidal waves of motion circling its circumference. If disturbed again, this time more sharply, it shivers, vibrates, and oscillates, and can suddenly come apart into two smaller spheres that drift away, still trembling. If the straw is pulled away, the strawberry drink floats freely in the cabin until air currents carry it toward the wall, ceiling, or floor. Once it touches any relatively flat surface, it changes its shape instantly from a sphere to a perfect hemisphere—looking now like a Jello dessert molded in a round mixing bowl and dumped onto a serving plate—and it stays attached for hours or days unless mopped up with a towel or, better yet, chased down by a thirsty astronaut.

The crew's personal cleaning is limited to rubdowns with wet wipes or washcloths; there is no shower on board. Hands can be washed and washcloths moistened by inserting them into a small domed device, mounted on the side of the galley, that shoots out a spray of water. Since most shuttle missions last only a week, a full-fledged shower facility—like the spray-and-suction shower that was used with only moderate success on Skylab—was deemed too space-consuming to be incorporated into the orbiter.

On the other hand, the shuttle's toilet facilities—what NASA euphemistically calls the waste-collection system—are at last a functional answer to the nagging problem of disposal of body wastes. The first astronauts had to urinate into strapped-on hoses and were rather

Norm Thaggard eats scrambled eggs on the ceiling of the mid-deck. He is joined for breakfast by Bob Crippen, commander of the second flight of the orbiter *Challenger*.

The author photographs Vance Brand shaving on *Columbia*'s mid-deck. His towel and washcloth float nearby.

ignominiously subjected to defecating into plastic bags, which were sealed, stowed, and returned to earth for analysis. Much to every astronaut's relief, the analysis is no longer considered necessary, and the shuttle orbiter is equipped with a rather conventional-looking toilet, which can accommodate both men and women, in its own tiny mid-deck compartment. Similar in most ways to water-flush units, the orbiter's toilet substitutes a powerful airflow for the force of gravity, pulling wastes into a sealed disposal chamber.

Sleeping is perhaps the one daily function that demands *less* design, planning, or deliberation in space than it does on earth. The eternal fall of orbit provides genuine "flotation support" that is unmatched by any mattress, and the weightless environment demands few special sleeping accouterments. The mid-deck bunks carried on most missions provide a measure of privacy and a means of keeping slumbering crew members out of the way of astronauts at work. Each narrow bunk contains a reading light, ventilation ducts, and a

Astronaut Bill Lenoir, suited for a spacewalk, watches a rubber shark that has gone to space—a souvenir for the divers at the underwater training facility back home.

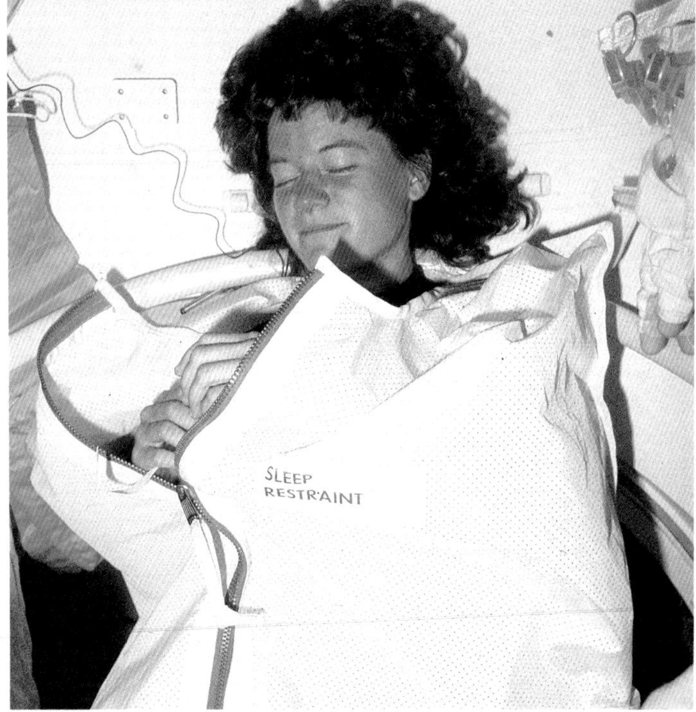

Astronaut Sally Ride sleeps in a "sleep restraint" inside *Challenger*. Mattresses are not needed in space, and astronauts tend to sleep wherever they can find a quiet corner.

PAYLOAD BAY FLOOD
RMS SPOTLIGHT
RMS CAMERA

The author, photographed by Vance Brand, watches the earth through an upper window on *Columbia*'s flight deck. A light meter floats in front of him.

sleeping bag attached to a padded board that holds the sleeper in place. But crew members often do not bother to use their bunks, particularly on missions with only a single work shift. They simply don eye masks and ear plugs, if needed, and hook themselves into an out-of-the-way corner, some sleeping fitfully, others falling asleep easily, sprawling in three dimensions instead of our earthbound two.

The comfort of sleeping in zero gravity has certain drawbacks, however. As on earth, it is quite possible for a hard-working crew member, now in the fifth day of a busy mission, to fall asleep in the middle of a somewhat tedious task. But, unlike on earth, the mission specialist is not warned that he is falling asleep by his head nodding over, nor does the checklist fall from his now-relaxed hand. His eyes simply close and he sleeps. Experienced crew members become aware of this potential embarrassment and quickly adopt a buddy system to avoid an over-long period of silence in response to a call from Mission Control asking how an experiment is progressing.

A simple exercise treadmill aboard the orbiter, designed by physician-astronaut William Thornton, has taken the place of the bulky and less versatile bicycle ergometer that was used for exercise on Skylab. To use the treadmill, the pilot straps on a belt and a shoulder harness that he attaches to the base by means of bungee cords. The base can be mounted on a floor, wall, or ceiling, and once the tension of the bungees is adjusted, the pilot simply runs in place against a movable track. He can jog across North America in about fifteen minutes, and it takes only ninety minutes to run around the world.

While all crew members are encouraged to exercise regularly in orbit, it is primarily the commander and pilot—who are responsible for bringing the orbiter safely back to earth—who spend daily periods on the treadmill on relatively short shuttle missions; regular running is presumed to be important to maintain mental and physical acuity for the critical deorbit and landing maneuvers. The remaining crew members tend to exercise as much or as little as their own preferences dictate.

While a few crew members are virtual marathoners in space, watching the earth roll by is the principal form of recreation for the majority of astronauts. During a span of only a week in space, the diaries and music that have been brought along somehow do not receive much attention. Mission assignments keep crew members at work as much as twelve hours a day, and after taking sleeping, eating, and housekeeping time into account, little leisure time remains. But this is time that the crew relishes, a chance to contemplate the amazing environment and circumstances, simply to enjoy the exquisite floating, or to play in it as if it were

The crew of the tenth shuttle mission celebrates the completion of the first experimental flight of the manned maneuvering unit. CLOCKWISE FROM TOP LEFT: The astronauts are Hoot Gibson, Vance Brand, Bob Stewart, Bruce McCandless, and Ron McNair.

a waterless, weightless swimming pool—astronauts as acrobats, turning flips and flying, playing weightless kinds of catch, blowing shimmering balls of water that float across the cabin.

During the long course of the eighty-four-day mission of Skylab 4, it was the absence of enough scheduled leisure time that became most irksome to the three-man crew. Astronauts Gerald Carr, William Pogue, and Edward Gibson became somewhat impatient early in the mission. They often complained about Skylab's poorly designed experiments, faulty equipment, awkward toilet facilities, poor lighting, drab clothing, and bland food. But it was the issue of time—too little time to perform experiments, to rest, and especially to do nothing at all—that seemed most bothersome. An open request from the crew for more leisure time was plainly not a part of NASA's flight plan, but mission controllers did agree to reschedule experimentation time for the remainder of the flight. But too much remained to be done, they reckoned, to give the crew whole days or hours off.

Finally exasperated at having virtually no free time between dinner and bedtime each night, Carr, the mission commander, tried to make his case again:

> The problem here is that right as of today none of us has as yet had any time to sit down and read or write or just stare out the window unless we do it after bedtime when the ground stops talking. . . . I think you'll find that you'll get better work out of us, we'll be more rested and more efficient if we . . . have some time for just plain quiet relaxation with nothing bugging us, no requirements on our time—just a period of time to be quiet.

Carr's comment certainly countered the popular notion that astronauts are somehow superhuman and can work efficiently around the clock without any ill effects. But his statement was blunt evidence of another way in which living in space does not seem to be very different from living on earth. As well as needing to work, eat, and sleep in a life-sustaining environment, people aloft in space for long durations seem to need to unwind, to give quiet consideration to who they are and to where out of the world they find themselves.

There are still too few shuttle missions launched into orbit, and each is still far too expensive and demands too much technical work to allow crew members big blocks of time for contemplation. Yet the time has come for bona fide passengers to join the astronaut crews, and for the first music to be composed and the first canvases painted from the compelling vantage point of empty space. As earth orbit becomes less foreign, its familiarity will make almost every aspect of human life seem appropriate to that environment.

The earth will doubtless be the home of the human species for centuries to come. This is not an era in which we have learned that we can live without the earth—space journeys only amplify our awareness of the importance of our planet, at least to us. It is, instead, a time in which we are beginning to discover that we can briefly and successfully travel away from the earth, that we can live, and live creatively, beyond it.

MANY MISSIONS

At Work in Space

During the first twenty years of manned American space flight, the business of space was the exploration of orbital flight itself. Our principal purpose in rocketing beyond the earth and in traveling to the moon was to explore a means of travel into a realm that was utterly new to human experience. The six Mercury flights of the early 1960s were designed to determine whether human beings and the machines that carried them could survive the pounding acceleration of liftoff, the bizarre phenomenon of weightlessness, and the fiery return to earth. The Gemini program—a succession of ten missions in 1965 and 1966—proved our ability to rendezvous and dock vehicles in space and offered evidence that astronauts could indeed live in weightlessness without dangerous physical consequences. Next, the eleven Apollo missions set out to prove—and did so spectacularly—that we could transport human beings to the moon, *to another world,* then return them safely to earth. The three Skylab flights of the mid-1970s tested the physical, mental, and emotional reactions of nine astronauts to prolonged periods in orbit, yielding evidence that humans can adapt remarkably well to the exotic conditions of space. Then in 1981, with the inaugural test flight of the Space Transportation System, we were able to prove that an aerospace vehicle could be launched like a rocket, maneuvered in orbit like a spacecraft, then flown through the atmosphere like an airplane to a safe landing on earth. And most importantly, this vehicle could be flown again and again. Unlike earlier spacecraft, it did not have to be retired to a museum after its maiden journey.

For more than two decades, the essential work of space travel focused on the technological refinement of the vehicles and support systems that could take us to space and back again, and on the study of the human response to those journeys. Still in that era of experimentation, the first four missions of the shuttle orbiter *Columbia* were test flights designed to gauge the Space Transportation System's capabilities. It was the flight of STS-5, launched

PRECEDING OVERLEAF: Astronauts Story Musgrave and Don Peterson perform the first experimental spacewalks of the shuttle era. Each wears a backpack that provides complete life-support, but each is tethered for safety to a line on the gunwale of the cargo bay. The gold foil-wrapped truss visible in the bay held a communications satellite that has been deployed.

OPPOSITE: A Gemini capsule has successfully docked with the Agena experimental docking target in this 1966 photograph. The Agena's communications antenna extends out from the craft like a radio aerial on a car.

A reaction-control thruster begins to fire on STS-6. A total of forty-four thrusters, or minirockets, are located in the orbiter's nose and tail. The thrusters fire in computer-controlled combination to maneuver the orbiter in space.

on November 11, 1982, that marked a fundamental shift in the goals of manned space flight from exploration to operation—from testing the means of getting into space to using the resources found there.

When the crew members of the fifth shuttle mission—Vance Brand, Bob Overmyer, Bill Lenoir, and Joe Allen, who dubbed themselves the "Ace Moving Company: Fast and Courteous Service"—deployed two communications satellites that had been carried to space in the orbiter's cargo bay, they initiated a new era in which the business of space flight became business itself. Today each shuttle mission is devoted to a wide variety of operations—from satellite deployment and repair to the manufacture of pharmaceutical compounds that are difficult to create on earth. A recent survey indicates that more than eighty companies around the world are interested in using the shuttle for a total of more than two hundred commercial and experimental projects. By 1990 a fleet of five or more orbiters may be flying twenty missions a year to fulfill this need.

The astronauts aboard each shuttle mission are no longer simply space explorers but, rather, a collection of skilled space workers—men and women whose jobs are to maneuver the craft in space, to unload new satellites being delivered to orbit, to pick up and repair or bring home satellites that have developed problems, to work outside the cabin in life-supporting pressure suits, or to perform experiments in a large laboratory carried aloft by the orbiter. The pilot-astronauts of the first decades of space flight have been joined by engineers, scientists, and technicians—known as mission and payload specialists—charged simply with getting the jobs done, with carrying out the crucial and often creative work of space.

The commander or pilot maneuvers the shuttle orbiter in its 17,500-mile-an-hour coast around the earth from either the forward cockpit or the aft crew station on the orbiter's flight deck. In the vacuum of space, the speeding craft can travel in any position—tail, wing, or belly forward—with equal ease, and crew members must periodically alter its *attitude,* the direction in which it points, to place it in proper alignment for the deployment of a satellite, to shield its windows from the debris of the satellite's rocket exhaust, to perform a wide array of experiments, or to give a camera or a communications antenna a better view of the stars or the spinning earth.

Except for the roles performed by the two large orbital maneuvering system engines—to thrust the orbiter into its specified circular orbit at the end of the launch sequence, to make major changes in the orbital path while aloft, and to slow the craft enough at the end of the mission to allow it to touch the upper edge of the atmosphere—the orbiting vehicle is maneuvered in space by the forty-four small multidirectional thrusters, or minirockets, of the reaction-control system. These are located in the orbiter's nose and tail. Six of the thrusters are vernier jets—small thrusters that slowly and quietly move the hundred-ton machine to new attitudes. The remaining thirty-eight thrusters are primary jets, which are *not* silent. These are used for more rapid rotations and for translation maneuvers—straight-line movements of the orbiter, forward or back, left or right, up or down. The forward primary thrusters sound like exploding cannons at thrust onset; and during their firing, jets of flame shoot out from the orbiter's nose. The aft primaries, 60 feet farther away, sound more like the *whump* of the launch of a mortar shell when they initially fire. The orbiter reacts to the primaries' shove by shaking slightly and by moving very noticeably. For the

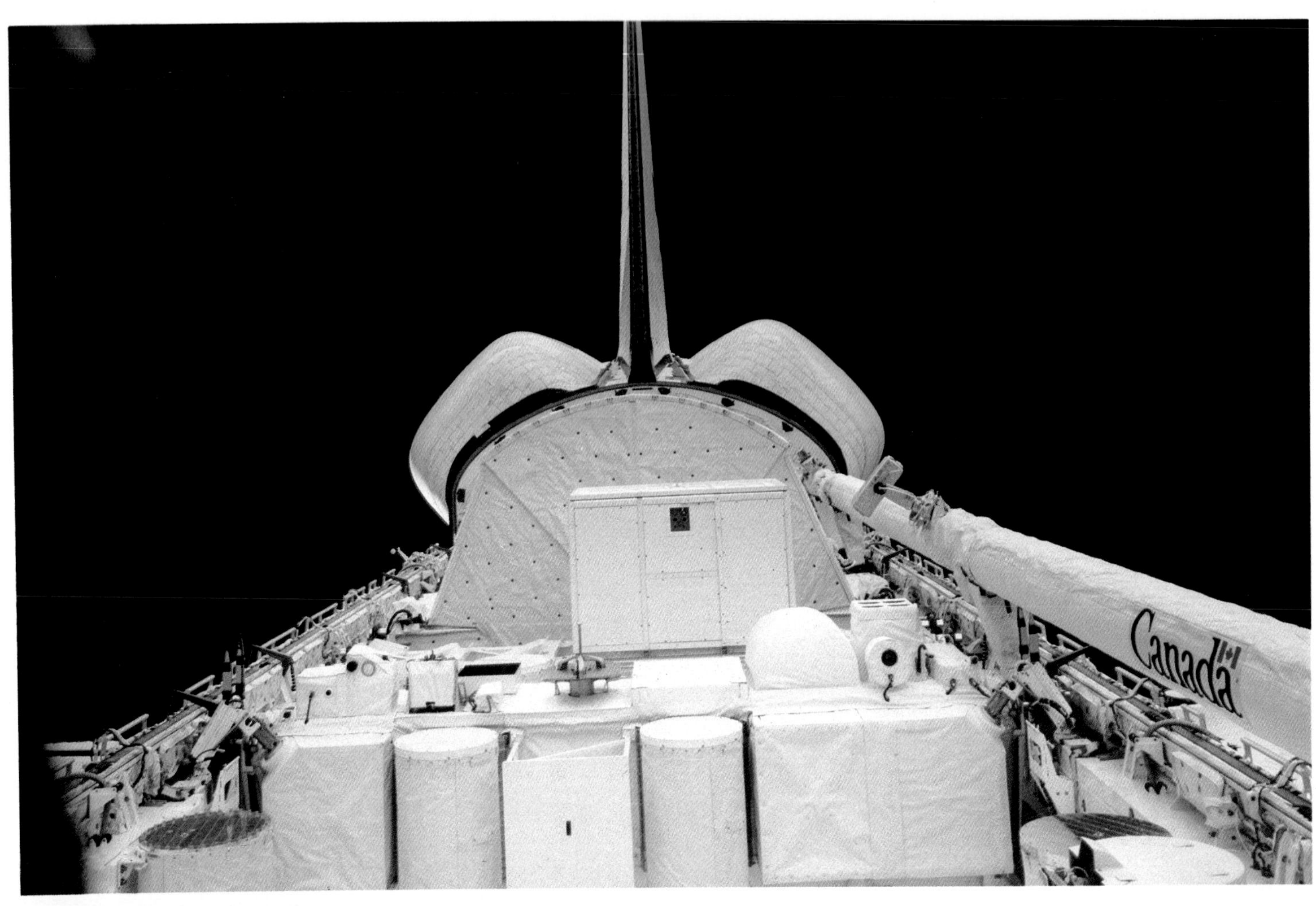
Canada

crew on board, a series of attitude changes using primaries resembles a World War I sea battle, with cannons and mortars firing, flashes of flame shooting in all directions, and the ship's shuddering and shaking in reaction to the salvos.

But for most maneuvers the vernier jets are sufficient, and to shift attitudes the pilot begins to move the orbiter into a 180-degree roll, for example, by manually rocking a rotational hand controller that is analogous to an airplane's control stick. The shuttle computers instantly translate the movement of the hand controller into commands to the necessary thrusters to fire in proper combination, and the orbiter smoothly, silently rolls wing over wing, its tail now pointing down at the earth, its black belly aimed toward the blackness of space.

The pilot or commander can similarly execute pitch and yaw rotations by twisting the hand controller in different directions and can, with the aid of another hand controller, execute translation maneuvers. Free of the constraints of the atmosphere, the craft can do more than fly; it can simply, and with elegant precision, turn or roll or go straight where its crew wants to take it, propelled by its rocket thrusters. Crew members strapped in seats or held lightly by toe loops or handholds move smoothly with the orbiter, of course, but one who is floating freely does not move as the spaceship moves around him, and he can watch with great delight as the ship maneuvers "without" him until one of the cabin bulkheads or attachments finally touches him. Correspondingly, the mission specialist, floating freely in the mid-deck and engrossed in making entries in her logbook, is surprised when she looks up to find that the ceiling has become a wall, and the wall is now a floor.

Although all flight operations can be controlled from the forward cockpit, the maneuvering controls mounted in the aft crew station, immediately behind the cockpit, are often used during the orbital phase of the mission. In fact, the crescent-shaped crew station—a compact array of maneuvering controls, radar displays, payload control panels, spacecraft systems monitors, computer keyboard, closed-circuit television screens, and two large windows that offer views into the cargo bay—forms a natural in-orbit control center for the entire craft. The crew can carry out all of the tasks that pertain to the payloads in the enormous cargo bay (except those that require spacewalks) while floating in the confines of the crew station.

The cargo bay itself occupies most of the orbiter's cylindrical fuselage. As big as a boxcar, it is 60 feet long and measures 15 feet in diameter, large enough to carry a bus to space, if ever there were a need, with room for an additional six or seven passenger cars.

OPPOSITE TOP: With the doors of the cargo bay open, the "Canadarm"—the remote manipulator system used to lift payloads in and out of the bay—is visible in its cradle on the bay's port gunwale.

OPPOSITE BOTTOM LEFT: In this photograph from STS-3, one of the bay doors is being closed. The doors are closed one at a time and must be properly sealed before the orbiter can begin its re-entry.

OPPOSITE BOTTOM RIGHT: Near the end of the fifth shuttle mission, the author photographs the interior of the cargo bay, lit with floodlights, after the bay doors have been closed. Visible are the Pac Man-like jaws of two sun shields that protected two satellites prior to their deployment.

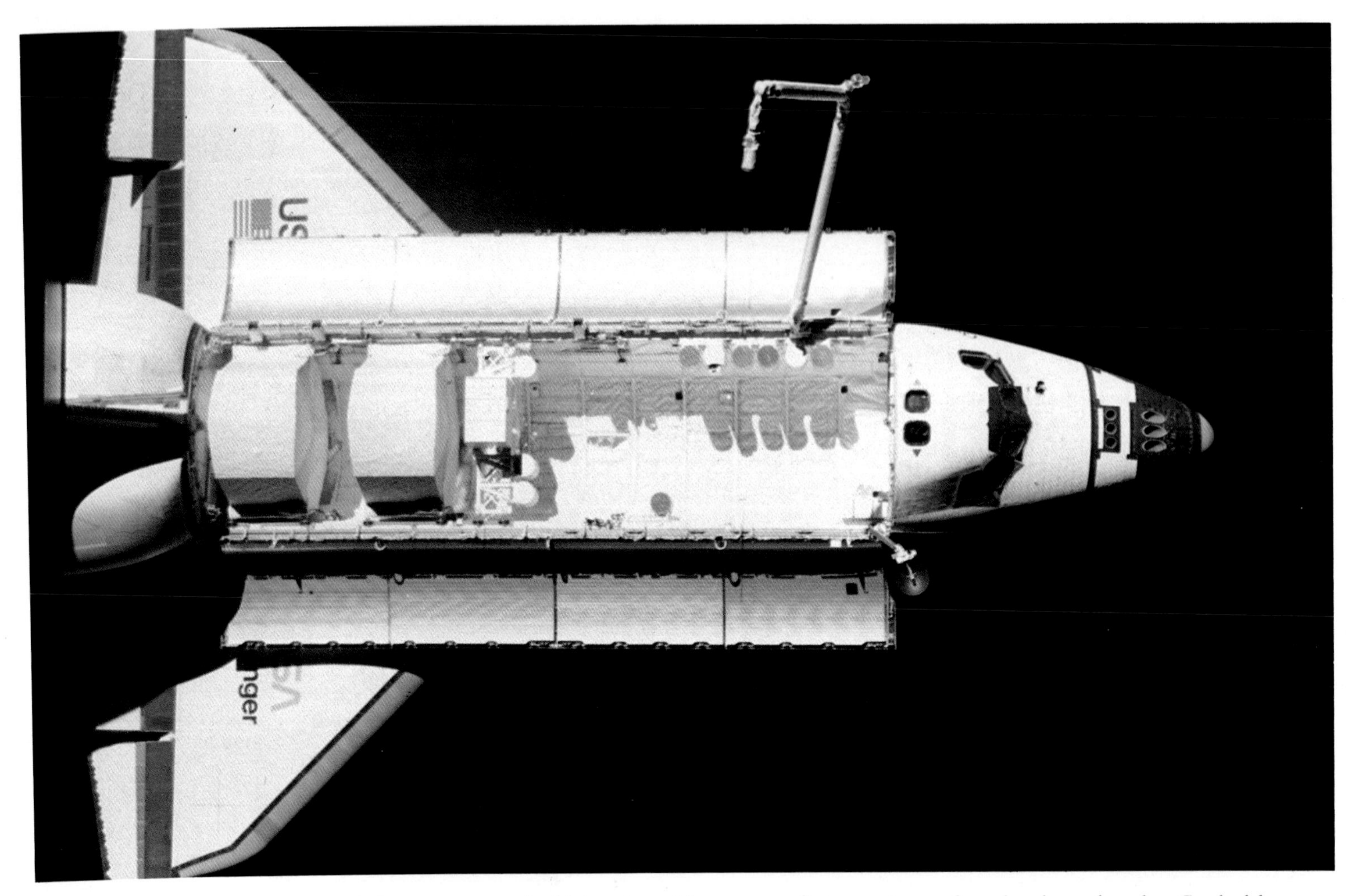

Challenger is photographed by an astronaut, about 300 feet from the orbiter.

OPPOSITE: Astronaut Bruce McCandless is moved toward a work area in the cargo bay by the arm of the orbiter's remote manipulator system, operated from the aft flight deck by Ron McNair.

This cavernous hold is adaptable to an infinite variety of payloads and tasks. Sealed by two overhead clamshell doors that run the length of the bay, it can accommodate on a single mission as many as five satellites—or a combination of satellites, instruments, self-contained experiments, and one or more components of the orbital laboratory called Spacelab. In addition, the undersides of the cargo-bay doors house important radiators that dissipate heat generated by the orbiter's electrical systems. Soon after the vehicle has entered its circular orbit, the pilot, floating at the aft crew-station controls, opens the bay doors and activates these radiators.

Floating beside the pilot at the aft crew station, the mission specialist, who is an expert in the operation of the *remote manipulator arm,* holds two hand controllers that govern the arm's movements. Also known as the Canadarm—it was designed and built by Toronto's Spar

McDONNELL
DOUGLAS
HUGHES
HUGHES AIRCRAFT COMPANY
WESTAR

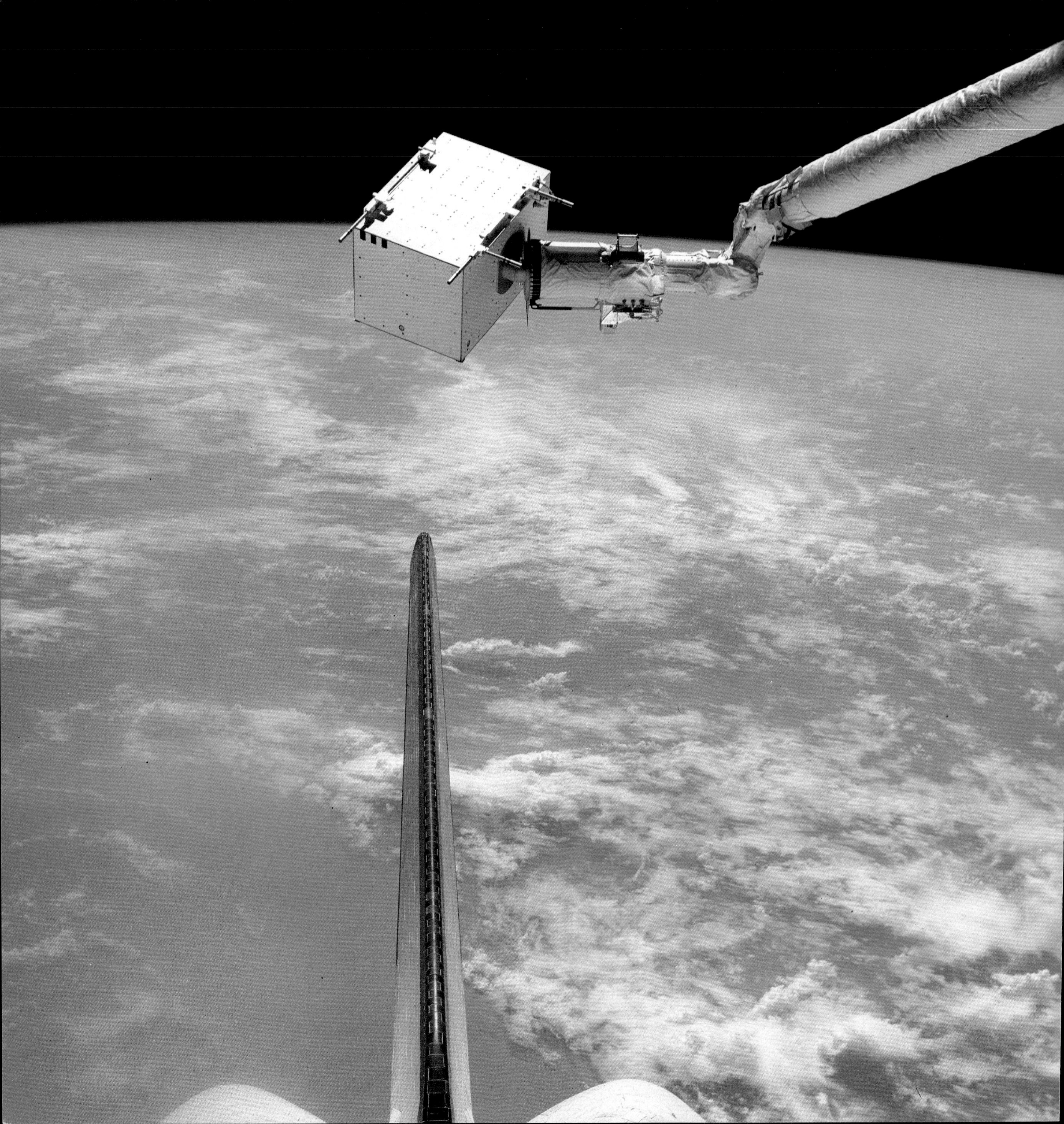

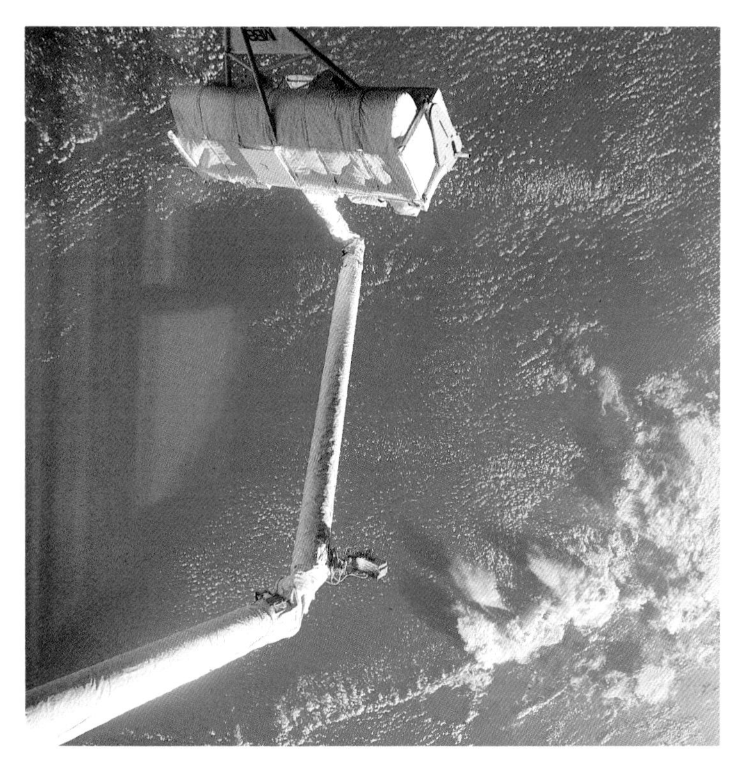

PRECEDING OVERLEAF:
The remote manipulator arm is used to grapple satellites in the cargo bay and to deploy them by lifting them overboard and releasing them. It can also snare damaged satellites.

Aerospace Limited—the remote manipulator is a three-jointed mechanical arm mounted near the bay's front bulkhead. It can lift payloads out of the cargo bay and release them overboard, as well as retrieve satellites or other objects and nestle them into the hold. The 50-foot-long arm is operated by electric motors at each of its three joints. Television cameras at its elbow and wrist joints send high-resolution pictures to a closed-circuit television screen on the flight deck and offer the mission specialist a clear, close-up view of its intricate and delicate movements. The hand of the arm, called an "end effector" in NASA's somewhat obscure parlance, is a three-wire capture mechanism that twists around a small probe, or grapple, attached to the satellite or a portable payload, then locks onto it, thereby enabling the arm to lift the weightless (but not mass-less) object out of its cradle and release it into its own independent orbit.

The deployment of communications, navigation, weather, mapping, and surveillance satellites is the nuts-and-bolts business of the shuttle orbiters. In the twenty-five years since the Soviet *Sputnik* first orbited the earth, satellites have revolutionized many aspects of life on earth, and the communications-satellite industry has grown into a multibillion-dollar-a-year business. An estimated 290 operable satellites of every type, owned by businesses and governments around the world, are currently orbiting in space, and the Space Transportation System alone is scheduled to deploy more than a hundred more during the remainder of the 1980s.

In its first few years of operational flight, the shuttle has emerged as a dependable, competitive alternative to the older method of satellite deployment—launching single satellites atop the expendable, one-time-only Delta, Atlas-Centaur, Titan, or French Ariane rockets. Fragile satellites that are carried into orbit in the shuttle's cargo bay can be heavier, contain more delicate machinery, and be less tightly packed than those that must endure a punishing launch in the nose cone of a rocket. Deployment costs for each satellite are reduced by carrying several satellites on a single mission. In addition, once these satellites are delivered, the orbiter can retrieve for repair other satellites already in low orbits, lengthening their working lives and further reducing costs.

The astronauts on any given shuttle mission are, in effect, the teamsters of its satellite-deployment operation—technicians who deliver the satellites into orbit, then deploy each one either by carefully lifting it overboard with the arm or simply by activating springs that gently nudge the satellite out of the cargo bay. The deployment of satellites destined for the

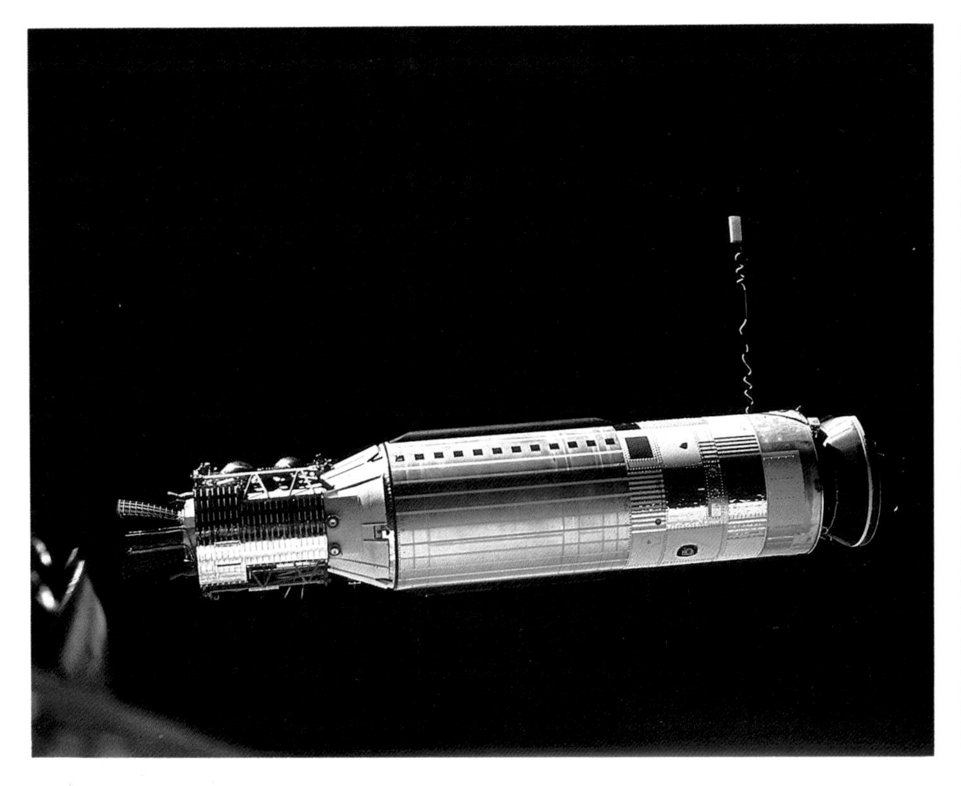

The Agenas, a series of unmanned docking targets launched atop Atlas boosters in 1966, provided docking for Gemini spacecraft. The ability to rendezvous and dock two vehicles in space would be an essential component of a successful landing on the moon and return to earth.

relatively low orbits where the shuttle normally flies—100 to 300 miles above the earth—is a rather straightforward operation. It involves releasing the satellite in its proper alignment from the manipulator arm—the newly independent satellite's orbital path and velocity now the same as those of the orbiter itself—then using the reaction control thrusters to maneuver the orbiter safely away and into a new orbit.

It is a more complicated enterprise to deploy a satellite that is destined for what is known as "geosynchronous," or "geostationary," orbit—the orbital path 22,300 miles above the earth's equator in which a single orbit takes precisely twenty-four hours to complete, causing a satellite to appear to hover motionless above a fixed point on the earth. As the mission specialists throw switches that power up the dormant satellite and begin to check that its systems are working properly, the pilot maneuvers the orbiter into its "deployment attitude"—a precisely calculated alignment based on the known position of the orbiter and the ultimate destination of its satellite cargo. A few minutes prior to the satellite's release from the cargo bay, the mission specialists draw back the Pac Man-like jaws of the insulating sun shield that surrounds the satellite. Then the satellite, mounted on a spin table that resembles an enormous record spin table, is "spun up" to fifty revolutions a minute. The spin gyrosco-

In December 1965, Gemini 6, carrying astronauts Wally Schirra and Tom Stafford, is able to maneuver to within a foot of Gemini 7, manned by Frank Borman and Jim Lovell.

pically stabilizes the satellite so that it will remain pointed in the proper direction after it is released. In spin tests on earth, the several-ton satellite and spin table roar as loudly as a subway train entering a station. In space, however, there is neither roar nor vibration. Not even a whisper travels from the spin table, now weightless on its bearings, through the orbiter's structure to the crew compartment. The mission specialists verify the satellite's proper spin rate on their instruments and continue to monitor the health of their expensive cargo, occasionally glancing through the windows to watch its silent and surrealistic whirl.

As the orbiter's path crosses the equator, the mission specialists fire pyrotechnic charges that release the clamps holding the satellite to the spin table. Four coiled springs that have been compressed between the satellite and the spin table now expand, and the satellite gently lifts out of its cradle, out of the sun shield, and away from the orbiter at the rather stately speed of 2½ feet per second, spinning silently into space.

Forty-five minutes later, after the pilot and commander have maneuvered the orbiter a safe distance away and turned its protective belly toward the satellite, a rocket motor at the base of the satellite automatically ignites with a bright plume of exhaust at the moment that the satellite again crosses the equator. The satellite then speeds away from the earth in an elongated elliptical orbit.

After as many as seventy hours and seven elliptical orbits (depending on the craft), the satellite reaches the geosynchronous position, where a second on-board motor fires to circularize its path around the earth. The satellite drifts toward its final "parking spot"; tiny thrusters adjust it to the precise geosynchronous speed of 6,876 miles an hour when it reaches its desired longitudinal location. Its antennas unfold, radio signals from its ground station move it into proper alignment, and the satellite is ready to begin as many as ten years of uninterrupted relay transmission.

The shuttle astronauts' role in the deployment of geosynchronous satellites is only an intermediary one, but it greatly reduces the chances that a satellite will be destroyed or its systems crippled en route from the earth to its remote destination. The deployment of a satellite from the cargo hold of the orbiter is as dramatic an event as the firing of a satellite atop a Titan booster from a Florida launch pad. But a satellite that is launched from the shuttle's orbit around the earth has already survived the hard part; the difficult, sometimes calamitous trip out of the atmosphere has been completed. Instead of the fountain of fire and barrage of noise that accompany a satellite launch on earth, an orbital launch is a silent, gentle, and slow-motion sendoff, beautiful in its choreography, stunning in its precision.

The retrieval of satellites or the repair of malfunctioning payloads in the orbiter's cargo bay can require one or two mission specialists to leave the security of the crew station and to go outside to make manual adjustments or to haul a satellite into and secure it in the hold.

Since astronaut Edward White first crawled through the open hatch of his Gemini capsule in June 1965 to become the first American to venture outside an orbiting craft, people have been fascinated by what are popularly known as "spacewalks"—the floating of astronauts in life-supporting suits, free of their homes in space, becoming a delicate sort of satellite themselves. But there is no *walking* to be done, of course; there is only the same strange weightless floating that occurs inside a spacecraft, and astronauts have always used the more prosaic acronym EVA—extravehicular activity—to describe the multifaceted business of "walking" in space.

The first Gemini EVAs were experimental, testing the reliability of the pressurized suits and the life-support systems that protected each astronaut, examining whether he could safely and efficiently accomplish a variety of tasks enveloped in his bulky and cumbersome

The earth-orbit rendezvous of Gemini 6 and Gemini 7 was the first in a series of experimental maneuvers that ultimately led to the rendezvous and docking in lunar orbit of a series of Apollo command modules with their lunar landers.

suit. By the time of the Apollo moon landings, EVA capabilities had been well tested, and on the surface of the moon astronauts actually could walk, or rather bound from place to place, pulled down by the moon's light gravitational tug.

During the shuttle era, EVAs continue to enhance human capabilities in space, enabling crews to undertake a variety of activities and repairs that simply cannot be accomplished by remote control by the manipulator arm. Although EVAs are not planned for every mission, each flight carries suits, equipment, and life-support materials for two-person EVAs, should they be required. Unlike the space suits of earlier eras, which were tailor-made for each astronaut, shuttle suits are available in a variety of sizes—several sizes of torso, many sizes of gloves, a variety of lengths of arms, legs, and feet, and a single-size bubble helmet. The shuttle suit is a two-piece garment—trousers with feet, and a hard-shell torso with arms and a backpack that contains life-support equipment. Although still cumbersome, this suit is easier and quicker to put on than the earlier one-piece models.

When two mission specialists prepare for a satellite-retrieval EVA, each dons an undergarment similar to a set of long underwear whose fabric is cross-hatched by liquid-filled

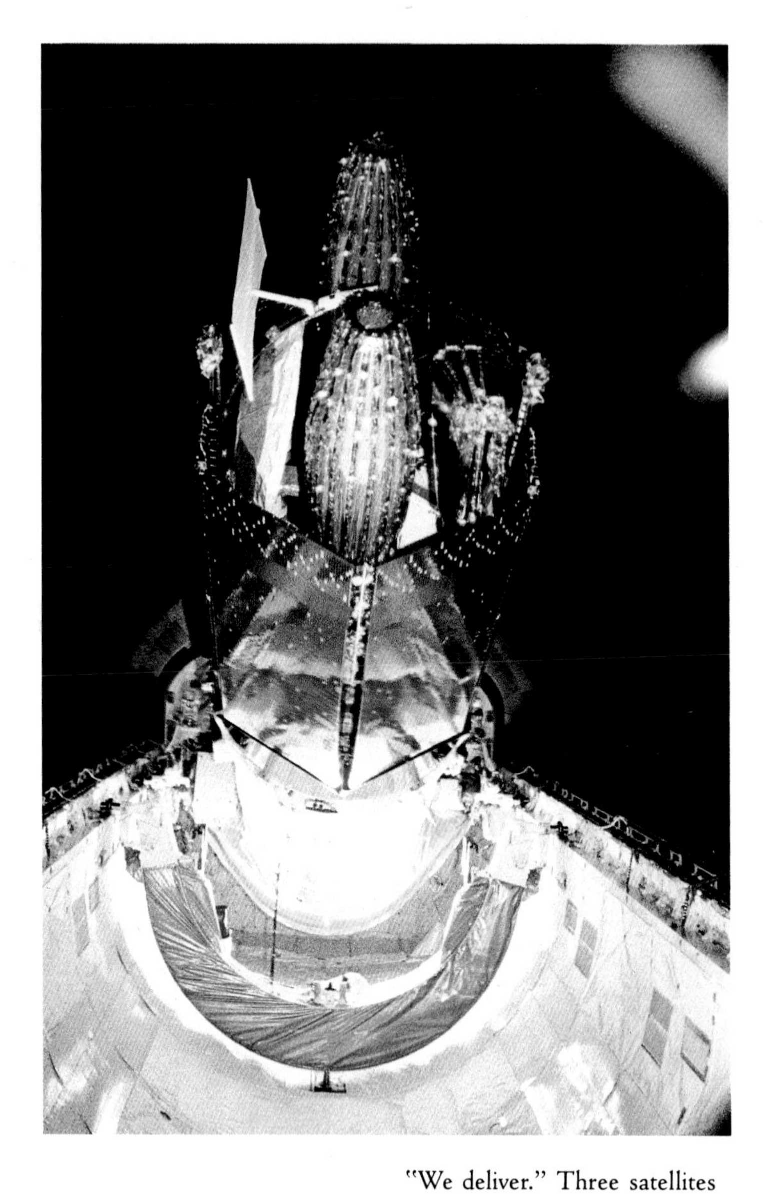

"We deliver." Three satellites are shown moments after deployment by shuttle crews. After the orbiter has maneuvered safely away, rockets aboard the satellites fire and send them toward their orbital destinations. ABOVE LEFT: A government-owned tracking and data relay satellite is deployed on STS-6. ABOVE RIGHT: A communications satellite that will serve India is launched on STS-8. OPPOSITE: The crew of STS-10 photographs the launch of a satellite, owned by Western Union, from the space shuttle in February 1984.

HUGHES
western union
Canad
MBB
SPAS-01

OPPOSITE: Owen Garriott checks equipment during a spacewalk on the second Skylab mission. Garriott is positioned near the center of a windmill pattern of solar panels that supply electricity for an array of instruments that study the sun.

OVERLEAF: Astronaut Bruce McCandless flies the manned maneuvering unit, a jet backpack that enables him to maneuver independently of the orbiter, on the tenth shuttle mission. Unassisted by umbilical lines or tethers, McCandless is literally a human satellite, having flown about 300 feet away from the orbiter *Columbia*.

FOLLOWING OVERLEAF: Bruce McCandless tests the operation of the manned maneuvering unit. Hand controllers on the arms of the unit operate the twenty-four gas thrusters that move the astronaut. A television camera is mounted on the top of McCandless's helmet. A still camera, wrapped in an insulating cloth, is mounted on the upper right of the unit.

cooling tubes, then enters the cylindrical air lock on the orbiter's mid-deck that connects the cabin to the cargo bay. The two crew members check the condition of both suits, then each begins to put one on—trousers first, torso next. Each connects the cooling umbilicals from the life-support backpack on the torso to the undergarment, then seals the two halves together with a pressure-tight metal ring. Through the operation of the controls on the torso of the suit, the liquid in the undergarment is heated or cooled in seconds to a temperature that the wearer finds comfortable. A lightweight cap that holds a microphone and earphones in place—known formally as a communications carrier, informally as a Snoopy helmet—is next, followed by gloves and finally the bubble helmet.

The two astronauts begin a 3-hour "prebreathing" period, during which the nitrogen in their bloodstreams is purged by breathing pure oxygen from the space suits' life-support systems, preventing the painful and potentially fatal condition known as dysbarism, or the bends, in which nitrogen bubbles collect in the body's joints when the air pressure (or water pressure, in the case of an undersea diver) around the body is reduced. Although the pressure in the suit is still at the normal atmospheric pressure of 14.7 pounds per square inch during the prebreathe period, it drops to about 4 pounds per square inch as the air inside the air lock is vented into the vacuum of space. The space-suit pressure remains at 4 pounds per square inch during the EVA, high enough to provide sufficient oxygen to breathe and adequate body pressure, but low enough to prevent the suit—in effect a body balloon—from being stiff or overinflated, which would hamper arm and leg movements.

The astronauts review plans for their EVA while they wait; then they take a short nap as they float inside the life-sustaining cocoons of their space suits. By the end of the prebreathe period, the normal operation of the suits has been verified and the air lock's inside hatch has been sealed and checked. Then the two crew members open the valve that discharges the remaining air in the air lock into space. They release the handle and pull open the outer hatch, an action accompanied by a hollow clang of metal against metal and a hiss as the last wisps of air escape into the vacuum. While still securely inside the air lock, each reaches outside to grasp a safety tether, pulls it in, and attaches it to a heavy ring on the suit. The tether is a thin lifeline the end ring of which slides along a line that runs the length of the bay, just as a dog's leash slides along a clothesline. Secured by their respective tethers, the pair floats through the open hatch in the front bulkhead of the cargo bay, pulling themselves along on a series of handholds mounted near the hinge line of the cargo-bay doors. Neither astronaut is connected to the orbiter by umbilical lines—each backpack provides complete

NASA

EV2
EVA PRESS

life-support and communications systems. Even though still attached by the thin tether, the astronauts can release their handholds and drift free, out of reach of the orbiter's gunwales. They can literally become human satellites, a thought that is both thrilling and somewhat sobering. Because there is nothing to push against and no air, however thin, to swim against, an EVA astronaut just a fingertip away from a handhold on the orbiter cannot move to it. By the laws of orbital mechanics, if the astronaut's orbit is not destined to intersect the orbit of the spaceship, no amount of patient and nervous waiting will help. Of course, a gentle tug on the tether is all that is needed to solve the problem. This simple action subtly changes the astronaut's orbital path (and also, very slightly, the path of the shuttle) so that the two paths now intersect, and the handhold can be grasped.

The view of the earth and space from the cargo bay is as spectacular and panoramic as the universe itself; it is no longer limited to what can be seen from the windows of the orbiter's cabin and is blocked only by the long fuselage. But the EVA schedule is too short to allow much time for awed sightseeing. In addition to retrieving the satellite, the two must transfer film cassettes and clean the lens of a camera mounted in the cargo bay, inspect and take readings from two experimental payloads, and retract and stow a portable antenna.

In order to make the satellite retrieval possible, the mission's commander and pilot—with the aid of the orbiter's on-board computers and tracking systems—have maneuvered the shuttle into the same orbit as the satellite, then pulled to within 300 feet of it. It is now the job of one mission specialist to go out to it, snare it, and bring it back. To reach the satellite, the mission specialist attaches herself to the manned maneuvering unit (MMU) that is stowed in the front of the cargo bay. Designed in true futuristic Buck Rogers style, the MMU resembles a backpack with armrests, or some kind of overstuffed rocket chair. The astronaut latches the hard-shell torso of her space suit to the MMU, disconnects her tether, unclamps the MMU from the bay's wall, then propels herself out of the bay by operating hand controls at the end of each armrest. As she manipulates the hand controls, control circuits within the unit send signals that fire small bursts of nitrogen gas from one or more of the twenty-four thrusters arranged around the MMU's exterior, propelling the astronaut—now completely free from the orbiter—into any rotational or translational movement she desires. The MMU is this spaceship's special dinghy, and the astronaut who operates it maneuvers away from the ship and out into the empty sea of space.

The MMU has enabled astronauts, for the first time, to orbit for brief periods without any umbilical line or safety tether to their spaceships, but use of the MMU is not considered

OPPOSITE: Astronaut Bob Stewart is attached to the manned maneuvering unit. Near his right elbow are tether hooks to which a variety of objects can be secured. A camera is mounted above the straps. The dark wands that curl in front of his helmet are light pipes that signal when the MMU's thrusters are firing. A checklist is mounted near Stewart's left elbow.

PAGE 114: Story Musgrave and Don Peterson perform a variety of experimental spacewalk activities on STS-6.

PAGE 115: Story Musgrave floats along handrails in *Columbia*'s cargo bay en route to the hatch that connects the bay with the pressurized crew quarters.

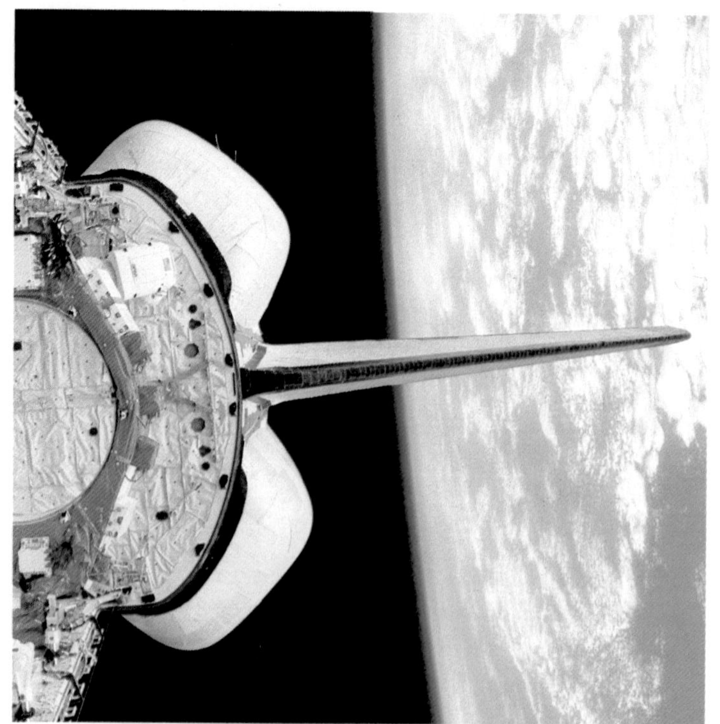

Inside the orbiter's cargo bay, Bruce McCandless tests a pneumatic tool that is used in satellite repair. The astronaut is held in place by a foot restraint on a platform attached to the end of the remote manipulator arm. The arm extends vertically beyond the top of the photograph. Visible on top of his helmet is a television transmitting antenna.

a special risk. The unit's redundant systems make it virtually fail-safe. But even if the unit were to fail somehow, the orbiter could maneuver to within 50 feet of a stranded astronaut, extend the manipulator arm, and bring the wayfarer safely aboard.

In this case, however, the unit operates perfectly; the mission specialist flies out to the nearby satellite, stabilizes its slow spin by grabbing a handhold, then flies back to within range of the manipulator arm with the satellite in tow. Operated by crew members inside the orbiter, the arm grapples onto the satellite and completes the job of bringing it into the bay. The astronaut flies the MMU back to its storage position while her partner checks to see that the new cargo is berthed and safely secured in its cradle. The two then work their way along the handholds to the open hatch of the air lock. They float through the hatch, unhook their tethers and return them to the storage locations outside, then seal the hatch behind them. The air lock is repressurized and this EVA is finished.

When the ninth shuttle mission was launched in November 1983, the cargo bay of the orbiter *Columbia* contained no satellites, nor was it destined to snare a crippled satellite already aloft and bring it back to earth for repair. *Columbia*'s cargo hold was filled for that flight with the first mission of Spacelab, a modular laboratory that allows scientists and technicians to observe and participate firsthand in a wide array of zero-gravity and high-vacuum experiments.

Spacelab is not what is popularly described as a space station; it is not an independent, self-sustaining, orbiting outpost that would be periodically resupplied with food, fuel, equipment, and personnel. Rather, Spacelab is dependent on the shuttle orbiter for its life support and electricity. It is securely bolted to the cargo bay and cannot be placed into its own orbit. Spacelab is, in effect, an extension of the orbiter's cabin, a versatile and sophisticated addition to its shirt-sleeve environment that provides scientists and technicians with a comfortable and relatively spacious working volume.

The laboratory, designed and built by the ten-nation European Space Agency, is composed of two main elements—the manned laboratory module and an external instrument platform, or pallet, on which can be mounted telescopes, antennas, or large experiments that require direct exposure to space. Each module is designed to be reflown as many as fifty times over a ten-year period. Depending on the experimental objectives of a particular mission, Spacelab can be outfitted with various combinations of lab modules and pallets to suit the experimenters' needs.

western union
WESTAR VI

2
NASA
MCDONNELL DOUGLAS
western union
HUGHES
HUGHES AIRCRAFT COMPANY
WESTAR VI

The six-man crew of this first Spacelab mission was launched in seats on the flight deck and mid-deck of the shuttle orbiter and prepared meals and slept in the orbiter's cabin as well. Yet as soon as the orbiter had entered its circular orbit around the earth, *Columbia*'s cargo-bay doors were opened to expose instruments on the pallets, Spacelab was powered up, and mission specialists Owen Garriott and Robert Parker and payload specialists Byron Lichtenberg and Ulf Merbold floated through the pressurized tunnel connecting the mid-deck to the lab module and promptly went to work.

While commander John Young and pilot Brewster Shaw managed the in-orbit operations of the shuttle craft during the nine-day mission, their fellow crewmen in Spacelab oversaw and actively took part in an ambitious total of seventy experiments dealing with a wide array of space-related investigations in the fields of biology, astronomy, earth observation, materials processing, and solar, atmospheric, and space-plasma physics.

Merbold, a German and the first European to fly aboard an American spacecraft, and Lichtenberg were the first payload specialists to fly on a shuttle mission. Although both men had received extensive training for their flight, their duties in orbit pertained specifically to the Spacelab experiments. They were indeed the first of a new breed of space workers—scientists and technicians who in the coming years will go to work in space in much the same way that they now work in laboratories on earth, yet taking advantage of the unique conditions and properties of their new environment.

Bruce McCandless flies the MMU above the cargo bay of the orbiter. Visible at the end of the MMU's arms is a docking apparatus that can be attached to a damaged satellite to maneuver it into the cargo bay for repair.

THE NEW OCEAN

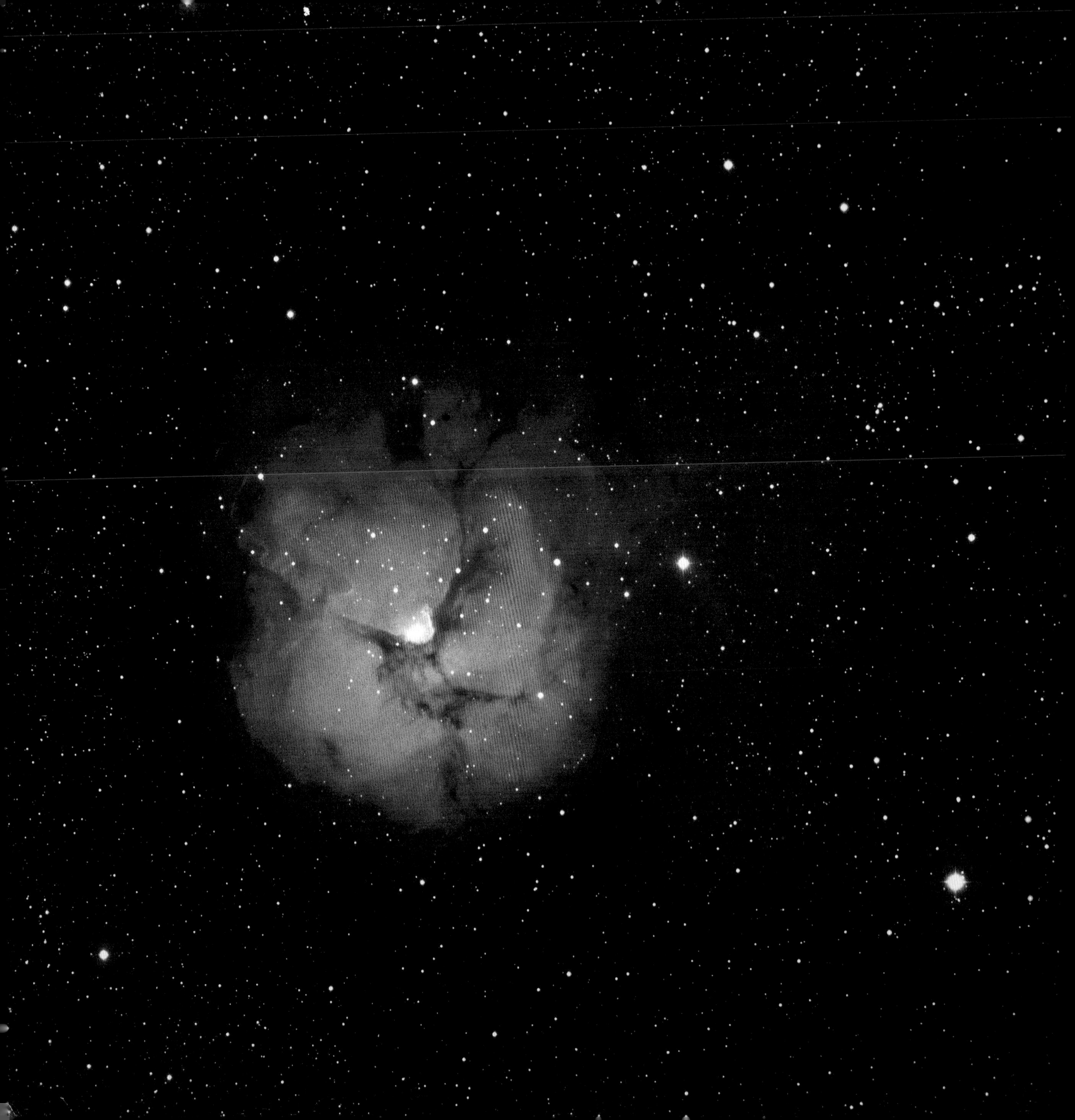

Journeys into the Solar System

When one of the fleet of shuttle orbiters is rocketed toward a 500-mile high orbit sometime in 1986, its payload bay will contain a 43-foot long, 13-ton telescope that is expected to revolutionize our understanding of our solar system, our galaxy, and the ever-expanding universe. Not since Galileo constructed the first optical telescope in the early seventeenth century, enabling him to discover the craters of the moon, the phases of Venus, sunspots, and the four largest moons of Jupiter, will humankind have made such an enormous single leap in its ability to observe the cosmos.

The Space Telescope, operating free of the light-obscuring and light-deflecting effects of the earth's atmosphere, will be able to gaze seven times farther into the universe than is now possible and to view the faintest light sources with a resolution that is ten times sharper than that of the best earth-based telescopes. It will operate in the visible range of light, as do earth telescopes, but it will also be able to observe light in the infrared and ultraviolet ranges, which the filtering screen of the atmosphere does not allow earth telescopes to do. The Space Telescope, the most powerful ever constructed, will expand the perceivable volume of the universe by an astonishing 350 times, and it will be of the utmost importance to our continuing efforts to understand the limitless realm of space.

The astronauts on the telescope-deployment mission will maneuver the orbiter into a precisely determined attitude before the manipulator arm grapples the tubular telescope and stands it upright in the cargo bay. Mission specialists, who will have received extensive training in the deployment and maintenance of the telescope, will power it up, check the operation of its various systems, and, if necessary, don pressure suits and go into the bay to make final adjustments before the arm lifts the telescope out of the hold and releases it. As the orbiter gently maneuvers away, the telescope's two power-producing solar panels will unfold, its antennas will be deployed, and, once in radio contact with its ground controllers,

PRECEDING OVERLEAF: The open star cluster Pleiades contains relatively young stars—not much more than one hundred million years old. The light of the stars in the cluster reflects off the dust cloud in which the cluster is embedded. (California Institute of Technology)

OPPOSITE: The Trifid Nebula belongs to the Sagittarius arm of the Milky Way. The red hues are produced by glowing hydrogen gas. The blue regions are primarily dust particles in the cloud reflecting light from stars in the nebula. (David Malin, Anglo-Australian Telescope Board)

it will be ready to begin its fifteen-year mission—looking into the remote reaches of the universe.

The telescope's guidance system will enable it to track and lock onto faint light sources with impressive accuracy. It will be able to fix its 8-foot-wide primary mirror on a specific target to within 0.01 arc seconds—an angle roughly the width of a dime in Washington, D.C., that is viewed from a telescope in Boston—and then stare at that object and collect light from it for up to ten hours. Light reflected from a nearby planet or generated by nearby suns or distant quasars will enter through the telescope's open front end and bounce off the primary mirror onto a smaller mirror. This second mirror will project the beam of collected starlight onto an array of sensors—imaging cameras, spectrum analyzers, and photometers—at the rear of the long tube. Information from the sensors will then be transmitted to a ground station where it will be processed and made available for study.

The Space Telescope will be able to operate around the clock, unimpeded by cloudy weather or by the obscuring effects of the earth's daylight skies. Nevertheless, officials at the Science Institute at Johns Hopkins University, which will manage the telescope's operation, predict that they will be able to accept only about a third of an anticipated 2,000 annual observation proposals from astronomers around the world. Yet those proposals that are accepted will hold enormous promise. Astronomers are hopeful that the Space Telescope will discover planets that are orbiting nearby stars such as Alpha Centauri or Sirius; none has ever been seen, but it seems probable that ours is not the only sun surrounded by a solar system. Tantalizing hints of a solar system around the star Vega were, in fact, suggested by the data obtained in 1983 by the earth-orbiting Infrared Astronomical Satellite.

The telescope will make crisp, close-up images of our neighboring planets commonplace. And it will offer astronomers abundant new information about multiple-star systems and about perplexing novae and even more mysterious quasars, billions of light-years away, whose light, emitted when the universe was in its infancy, is only now reaching the earth. Some astronomers and astrophysicists are confident that the telescope will help answer questions about the expansion of the universe and whether it will continue racing outward forever.

During the course of its fifteen-year operation, the telescope will occasionally be serviced and repaired on regular shuttle missions. Crews will fly their orbiters to within the manipulator arm's reach of the telescope. The arm will grasp the telescope and lift it into the cargo bay, where astronaut-technicians will replace instrument packages, adjust its support sys-

The command module *Gumdrop* is photographed from the lunar landing vehicle from which it has separated during the Apollo 9 mission. This experimental mission in March 1969 was designed to test the machinery that would soon venture to the surface of the moon.

tems, or perhaps install special optical sensors not yet devised. Should the telescope ever encounter major problems, its solar panels and antennas can be folded and it can be secured in the bay for a return to earth, repair, and a subsequent relaunch.

It was probably inevitable that humankind's ventures into space, which began with the orbit of a metal basketball-sized satellite called *Sputnik* in 1957, would not be limited to the near-space that surrounds the earth. We would yearn to go beyond the orbit of our own planet, to look at other worlds, and to visit them. For hundreds of years, ground-based telescopes had made Mars seem enticing and not so very distant; Venus appeared enigmatic, its mystery only heightened by the clouds that continually shroud it; and the moon, our own bright and seemingly barren satellite, was only 260,000 miles away—incredibly close by the standards of the solar system and crystal clear in even the smallest telescope.

In May 1961, President John F. Kennedy announced that the United States was ready to reach out to another world:

> I believe this nation should commit itself to achieving the goal, before this decade is out, of landing a man on the moon and returning him safely to the earth. No single space project in this period will be more impressive to mankind, or more important for the long-range exploration of space; and none will be so difficult or expensive to accomplish.

It is easy to forget how improbable Kennedy's bold challenge seemed at the time. Alan Shepard's suborbital flight had preceded the president's speech by a few days, but the United States was still almost a year away from its first manned orbit of the earth. The journey to the moon would demand the dedicated and precisely coordinated efforts of 300,000 people around the country; it would test the limits of our scientific understanding and of our technological capabilities; it would require the selection of dozens of astronauts with special skills in a variety of fields; and it would cost an almost inconceivable $25 billion. Even with that kind of national commitment, the ability of the United States to achieve its lunar landing goal by the end of the 1960s remained problematic at best.

But eight years and two months after President Kennedy first exhorted us to go forward, astronauts Neil Armstrong and Edwin Aldrin climbed down the ladder of their lunar module, *Eagle,* and walked on the surface of the moon. An estimated 600 million people around

TOP: The test lunar lander *Spider,* photographed from its command module in March 1969. BOTTOM: *Eagle,* the first manned craft to land on the moon, is photographed from the command module *Columbia* by astronaut Mike Collins in July 1969.

PAGE 128: The moon, the earth's only natural satellite, is photographed from the command module *Columbia* en route to the first manned lunar landing in July 1969.

PAGE 129: The moon, seen through the earth's hazy atmosphere, is photographed in 1973 from Skylab, the earth-orbiting laboratory.

PAGE 130: An Apollo command module is photographed against the black void of space. Communications antennas extend out from its shell, and its docking port is visible at its center.

PAGE 131: Astronaut Mike Collins photographs the far side of the moon from the lunar orbit of Apollo 11. The far side, perpetually hidden from view from the earth, has a much rougher surface than the near side.

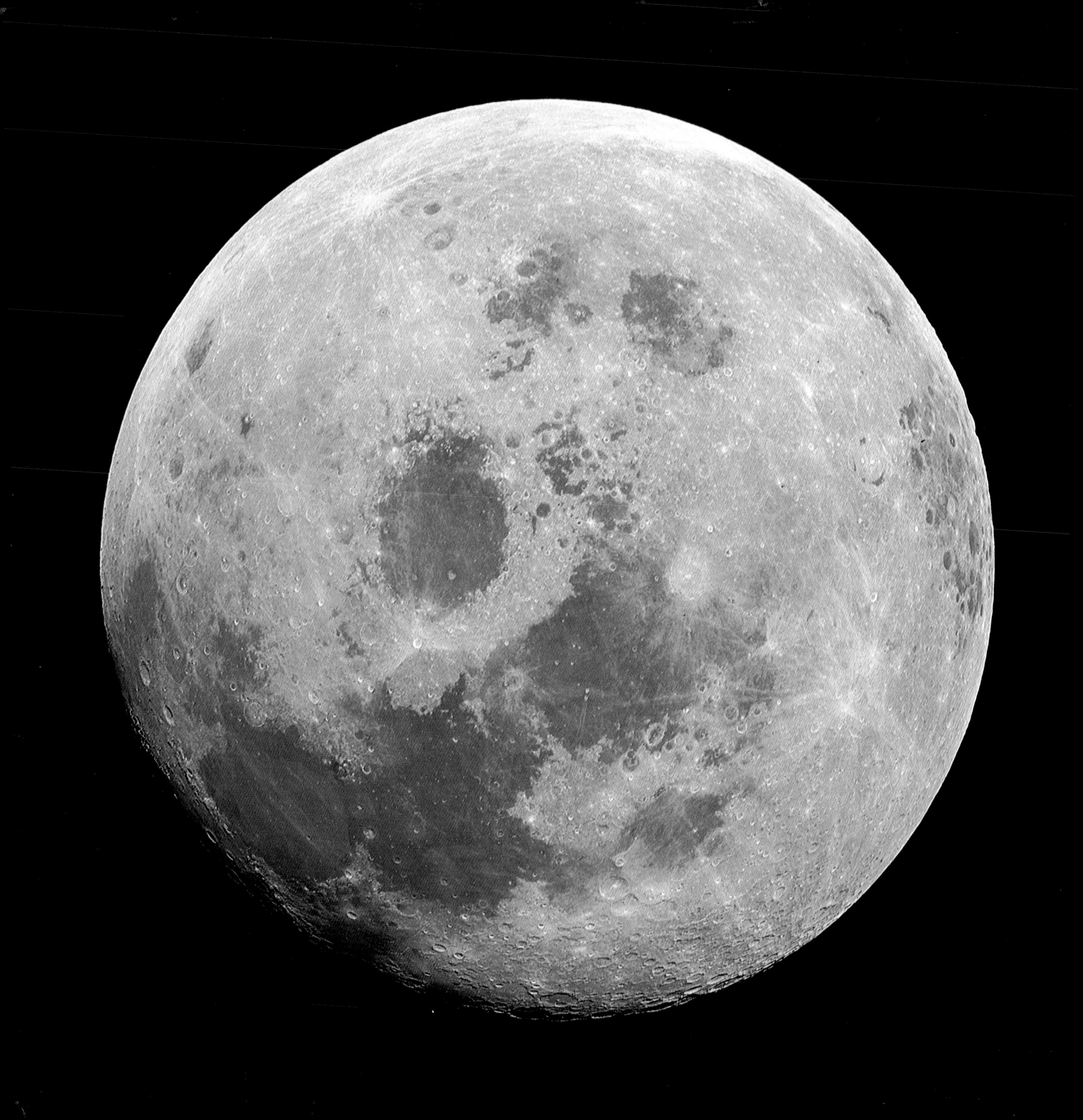

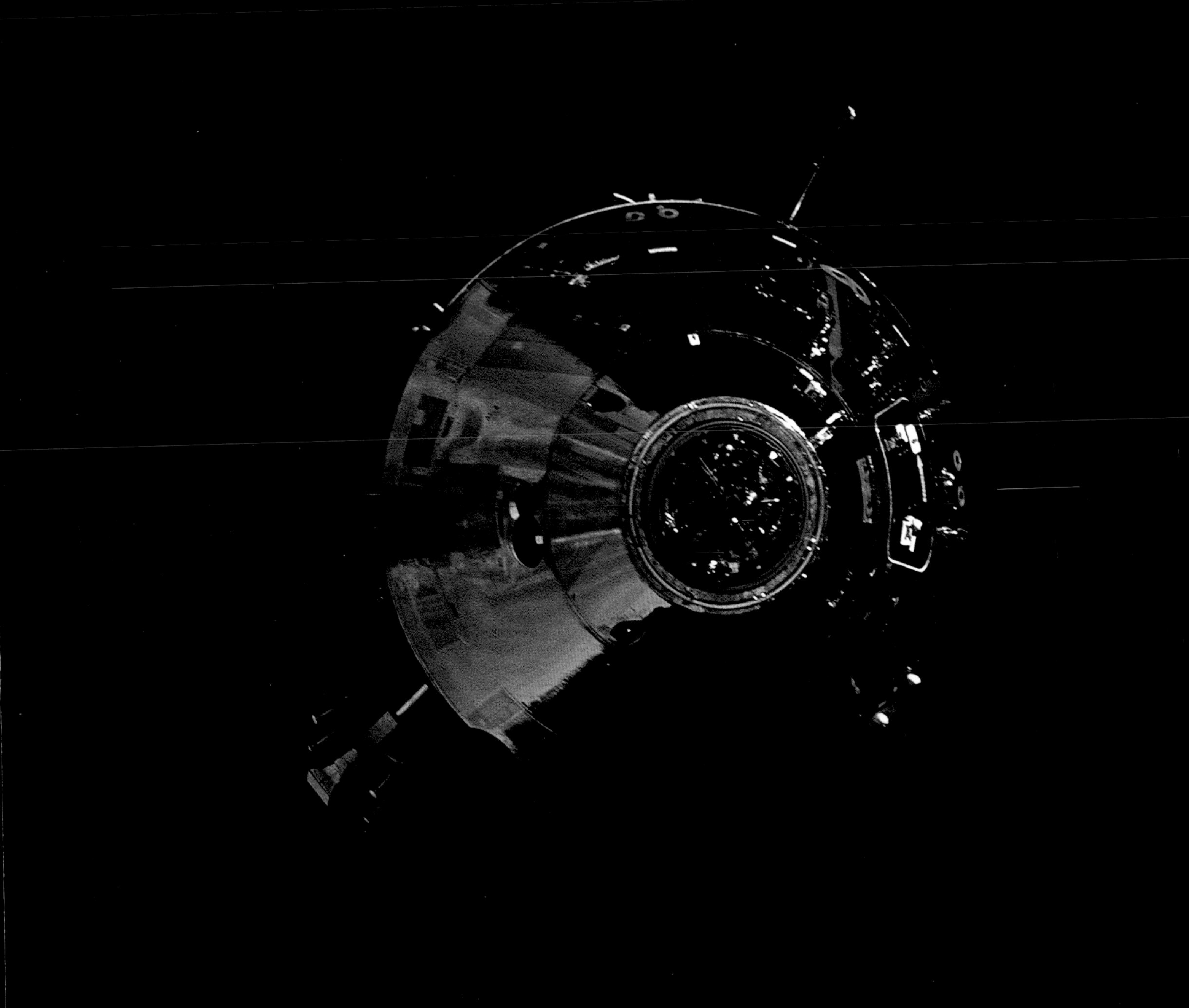

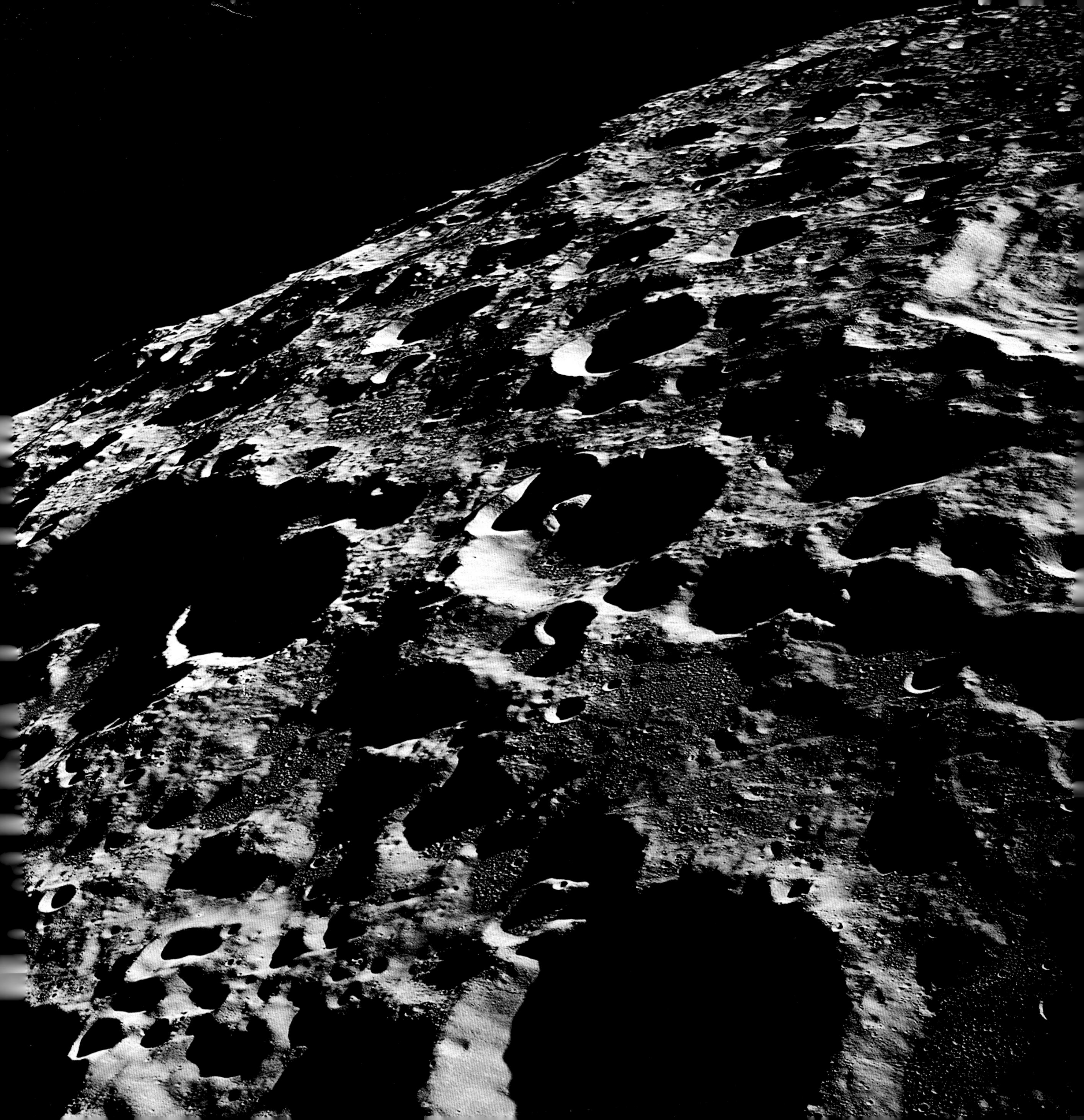

the world watched them on live television. Project Apollo, which would include six more missions to the moon, would be a great success. The huge three-stage Saturn rocket—363 feet tall (twice the height of the shuttle) and 6 million pounds—proved capable of hurtling the Apollo command and lunar modules out of the earth's grasp to reach the moon four days after liftoff. Exploration of the moon and the means of reaching it was the fundamental purpose of the Apollo program, along with the exercise of national pride and ambition. The lasting rewards from Apollo include new insights into the evolution of the earth, the moon's relation to its companion planet, and the early history of the solar system. And the new technologies that were devised to make the moon missions possible have been beneficially applied to virtually every facet of life on earth. But perhaps the most profound and unanticipated return from the lunar journeys has been humankind's new awareness of our beautiful and delicate planet.

We will look, and wonder, and ultimately travel to the moon again, to Mars, and to the Galilean moons of Jupiter because at our most fundamental levels we humans are travelers, investigators, wonderers. In the early 1960s, when the first lunar landing was still a dream, astronaut Walter Schirra, one of the first seven American astronauts, told Italian journalist Oriana Fallaci that the point of going to the moon was to prove that we could go beyond it:

> I know enough about the moon to know how unpleasant and inhospitable it is. . . . I know enough about Mars to know that you can't live there, you can't settle it. Mars and the moon are two ugly islands. So then, you say, what's the point of going to them? The point is to be able to say I've been there, I've set foot on them, and I can go further to look for beautiful islands.

The search for beautiful islands in the dark sea of the solar system has endured for as long as people have stood erect on earth, stared into the night sky, and dreamed what it would be like to travel there. Among the millions of stars that seem eternally fixed in discernible constellations, five bright wandering stars have long appeared unique and particularly compelling. The Babylonians, Greeks, and Romans gave each of these stars the name of a deity in order to describe its characteristics. The Romans called the swiftest one Mercury, after the messenger of the gods; the brightest was dubbed Venus, in honor of the goddess of beauty; the red one was named Mars, after the god of war; Jupiter, the king of the gods, gave his name to the largest and stateliest; and the slowest and faintest of the five was called Saturn, after the god of time.

OPPOSITE: At earth rise, astronaut Mike Collins photographs the ascent stage of the lunar lander *Eagle* as it returns to the Apollo 11 command module. Aboard *Eagle* are astronauts Neil Armstrong and Buzz Aldrin, the first two people to explore the surface of the moon.

PAGE 134, CLOCKWISE FROM TOP LEFT: Astronaut Buzz Aldrin descends *Eagle*'s ladder to the moon's surface; *Antares,* the Apollo 14 lunar lander, is photographed at Fra Mauro on the moon at sunrise; Aldrin sets up an experimental station near the lander *Eagle;* Aldrin unfurls a foil sheet that will collect particles of the solar wind for study.

PAGE 135: Astronaut Jim Irwin, photographed by Dave Scott on Apollo 15, stands beside the lunar rover that carried the two on explorations of the Hadley-Apennine region of the moon. The umbrella-shaped antenna mounted on the rover beamed television and voice communications to the earth.

UNITED
STATES

LEFT: Astronaut Gene Cernan photographs Jack Schmitt beside a lunar boulder dubbed "Splitrock" in the Valley of Taurus-Littrow. Cernan and Schmitt were the last astronauts to explore the surface of the moon.

ABOVE: The earth is visible above lunar lander *Eagle*.

But not until 1543, when Nicholas Copernicus, a Polish priest and scientist, forcefully theorized that the earth revolved around the sun—shattering the ancient notion that the earth was the center of the universe—did a plausible and simple explanation for the motions of the five special stars seem apparent. They too must have been revolving around the sun and reflecting its light, suggesting that they might be worlds much like the earth. The age of telescopic investigation of the planets, which began sixty years later with Galileo, proved over the next four centuries that nine planets orbit the sun, some of them orbited in turn by one, two, or a dozen or more "moons"—named after the earth's own satellite—which share the basic characteristics of the planets.

In the short span of two decades since the first of the Ranger series of unmanned lunar spacecraft initiated the era of interplanetary flight in 1962, we have flown by, orbited, or landed robots on more than forty spherical worlds, each one caught in a grand and graceful orbit of our sun. With the aid of long-lived and tenacious spacecraft containing an array of self-sustaining sensing instruments, cameras, and robotic devices, we have been able to travel to these worlds to explore and to learn. The distances between us and our neighbors in the solar system, with the exception of our own moon, are so great that they take nearly a year, two years, or up to ten years to traverse. They are distances that are still too great to permit manned exploration, but our dauntless machines now enable us to explore the solar system with our intellects and even our eyes, as photographs relayed to the earth so spectacularly prove.

When the Mariner 10 spacecraft flew by Mercury in 1974, for example, it relayed fascinating visual evidence that that planet very much resembles our moon. Like the moon, Mercury has no atmosphere. It is roughly the moon's size, and it too has a heavily cratered surface that, planetologists speculate, underwent the same bombardment billions of years ago as the moon did—a bombardment whose effects have long since been erased from the surfaces of the solar system's more volatile and turbulent planets.

Venus is very nearly the earth's twin in size, mass, and chemical composition. However, the U.S. Pioneer 1 and Mariner 9 and the Soviet Venera series of spacecraft discovered—over the course of a ten-year span of fly-bys, orbits, and landings that began in 1967—that that planet's mysterious cloud cover is a dense carbon dioxide atmosphere, laced with corrosive acids, whose surface pressure is more than ninety times the pressure of the earth's atmosphere at sea level. High winds race around the planet at 200 miles an hour, circling it once every four earth-days. The surface of Venus, mapped by radar and briefly investigated

OPPOSITE: The Apollo 17 lander *Challenger* sits in the Valley of Taurus-Littrow.

OVERLEAF: Mars is photographed by Viking 2 as it approaches the red planet in August 1976. At left is Ascreaus Mons, one of the giant Martian volcanoes. Ice clouds stream from its western flank. In the middle is the great rift canyon called Valles Marineris, and at right is the large, icy crater basin called Argyre, located near the planet's south pole.

by Venera landers, is flat and incredibly hot—about 900° F. Cameras and instruments aboard two Soviet landers were able to record Venus's forbidding surface characteristics for only a few hours before being destroyed by the heat.

The planet that had always seemed the most favorable and hospitable choice for human exploration was Mars. And, indeed, following the Mariner series of fly-bys and orbits of Mars in the late 1960s and early 1970s, the United States embarked on its ambitious Viking program—two separate spacecraft launched in the summer of 1975. The program was designed to map virtually all of the red planet and to investigate in detail whether Mars is home to primitive or even complex life forms. After nearly a year en route from earth, Viking 1 went into orbit around Mars on June 19, 1976, and its lander touched down on Mars's Plains of Gold on July 20, seven years to the day after the Apollo 11 lander brought the first human beings to the surface of the moon. The second Viking craft, also comprising

an orbiter and lander, arrived a few days later, its lander coming to rest on the Utopian Plains, halfway around Mars.

During the years when the Viking orbiters and landers operated, 97 percent of Mars was mapped far more precisely than had ever been possible. The orbiters observed enormous, fog-filled rift valleys, some of them stretching the width of the United States, and mountain peaks higher than Everest from which frail and feathery clouds trailed. They saw the geological evidence of great floods and liquid-caused erosion on a planet now so dry that swirling dust storms turn its thin skies a bright and gritty orange. The landers measured surface temperatures that ranged between a winter low of -191° F and a summer high of -24° F. And their robotic sample arms were able to scoop up Martian soil, which a variety of on-board instruments examined for evidence of life—a procedure that was directed from the earth *50 million* miles away.

This wide-angle composite photograph, taken from the surface of Mars by the Viking 2 lander, stretches nearly 200 degrees. The rusty color of the ground is limonite, hydrated iron oxide, which coats both the powdery soil and the rocks. The housing for Viking 2's soil sampler is visible at left, and the antenna that receives commands from the earth is visible at far right. The horizon in this photograph is about 2 miles from the spacecraft.

Jupiter's moons Io (visible in front of the planet) and Callisto orbit the largest planet in the solar system—a dense gaseous world that has no solid surface. This photograph was taken in March 1979 from the Voyager 1 spacecraft, one of two Voyager craft that were launched in 1977 to explore the outer planets of the solar system.

OPPOSITE: Jupiter's mysterious Great Red Spot is thought to be a centuries-old storm. Scientists believe its distinctive color may be caused by gases drawn up from deep within the planet. The Great Red Spot is large enough to contain three earths.

Neither Viking lander relayed evidence back to earth of any trace of life in its samples of Martian soil, though the soil appears capable of undergoing chemical change. But even if Mars did seem to be a dead planet, there was also something familiar about it. The exquisite images produced by the Viking orbiters and landers showed a world of rocky plateaus, red soils, and stark, barren canyons that resemble the high desert terrain of the American Southwest. Unlike Venus, Mercury, or even our own moon, Mars seemed recognizable, somehow homelike. And when the Viking 1 lander, the last of the Viking craft still operable, suffered a catastrophic failure on November 13, 1982, six years after it arrived, it nonetheless seemed certain that we would someday visit Mars again.

March 5, 1979, was a spectacular day in the history of science. During a span of about thirty hours that began on that day, an American spacecraft, Voyager 1, flew to within a few hundred thousand miles of the enormous planet Jupiter, larger than the rest of the planets in the solar system combined. Two television cameras aboard the spacecraft returned to the earth astonishing pictures, including proof of the previously unknown ring that surrounds Jupiter. The cameras observed Jupiter's puzzling

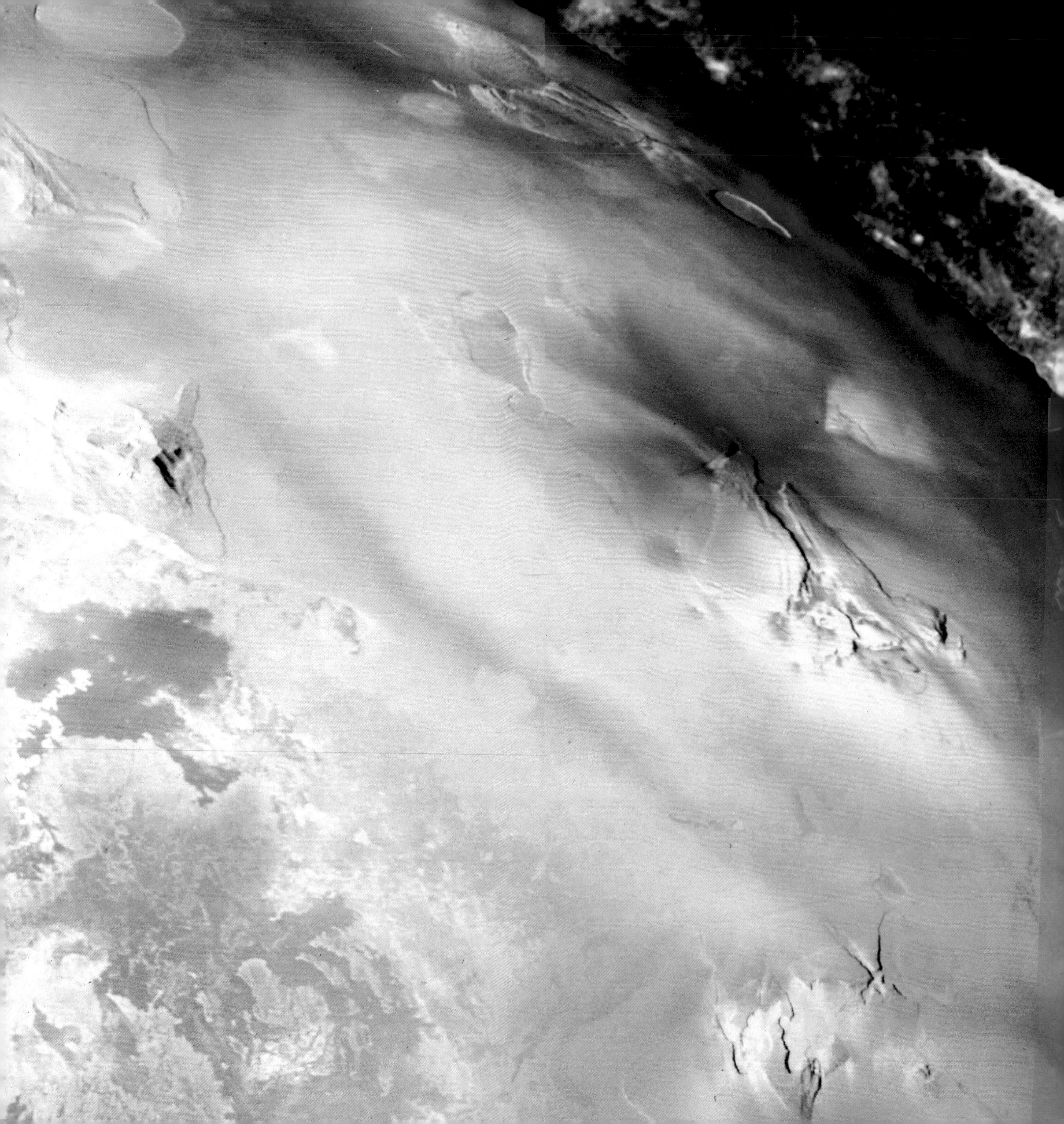

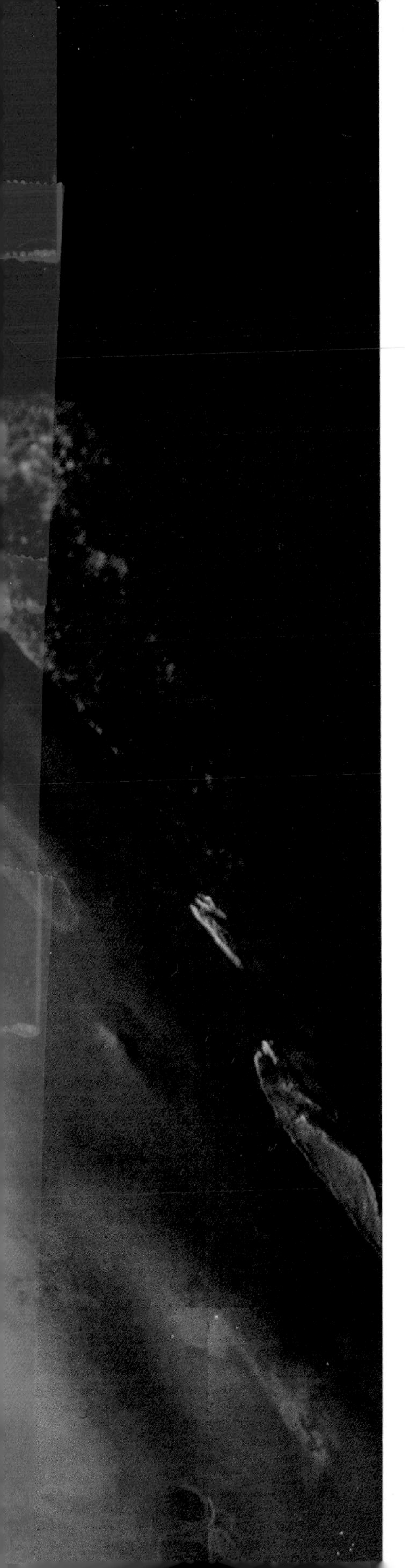

LEFT: Voyager 1 photographed the volcano named Pele in eruption on Io. The volcano's plume ascends to a height of about 175 miles, visible above the horizon in the upper right. When Voyager 2 arrived, four months after Voyager 1, Pele was inactive.

ABOVE: The orbit of Io takes it over Jupiter's Great Red Spot. Europa is visible to the right.

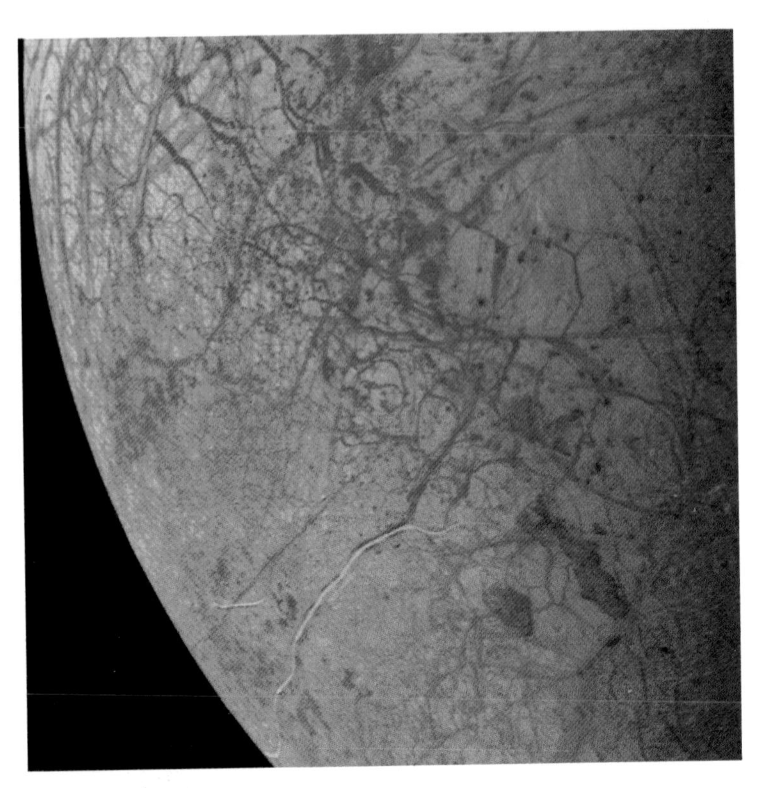

Jupiter's four largest moons were examined by the two Voyager craft, and these photographs were returned to earth. LEFT: Ganymede, the largest of the four, is the largest and brightest moon in the solar system. It reflects so much light probably because its surface is covered with a thick layer of ice. RIGHT: Europa, also ice-covered, is etched by blood-red fractures that may be filled with soil-like material from its interior.

Great Red Spot, wider than the earth and presumed by many planetologists to be a centuries-old storm. They photographed Jupiter's vast, swirling, gaseous surface (the planet does not appear to contain any solid material), and they returned hundreds of exquisite photographs of three of its four "Galilean moons": Ganymede, the largest, brightest moon in the solar system; Callisto, its icy twin; and Io, where to everyone's astonishment, the explosive eruption of a giant volcano was in progress. The spacecraft performed almost flawlessly; its systems were powered by a tiny pellet of plutonium because it was too far from the sun to make solar power feasible. It relayed information from its sensors and from the computer-generated dot-matrix system that composed each photographic image over a span of almost half a billion miles to NASA's Jet Propulsion Laboratory in Pasadena, California, where scientists and technicians anxiously began years-long studies of the new sights of and insights into Jupiter.

Voyager 1 was launched from Cape Canaveral in the spring of 1977. Four months later, it was followed by its companion craft, Voyager 2. The Voyager program, designed to visit and investigate Jupiter, Saturn, Uranus, and Neptune over the span of more than a decade, has been the most ambitious and potentially rewarding series of planetary probes ever

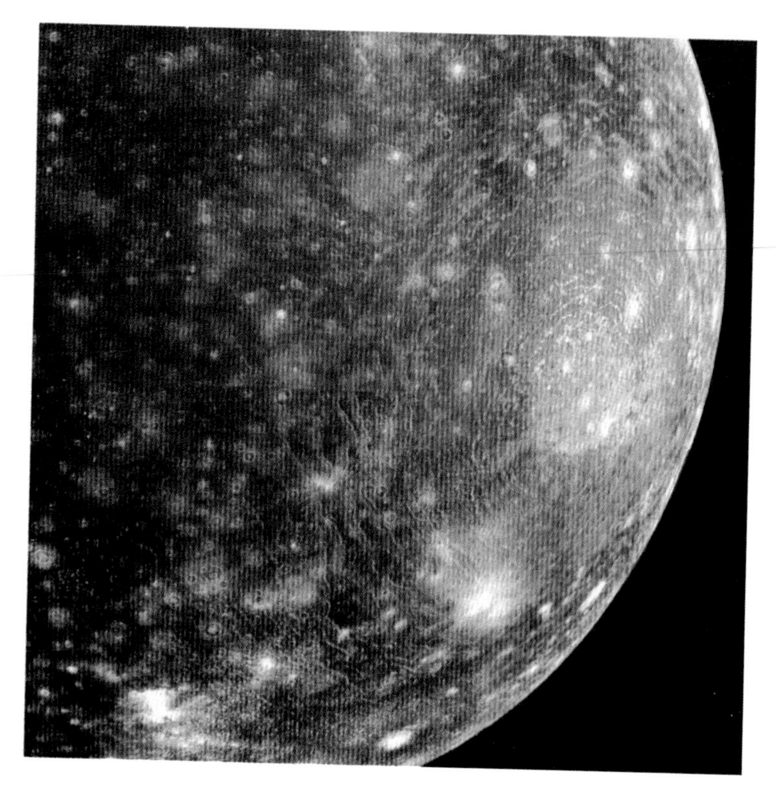

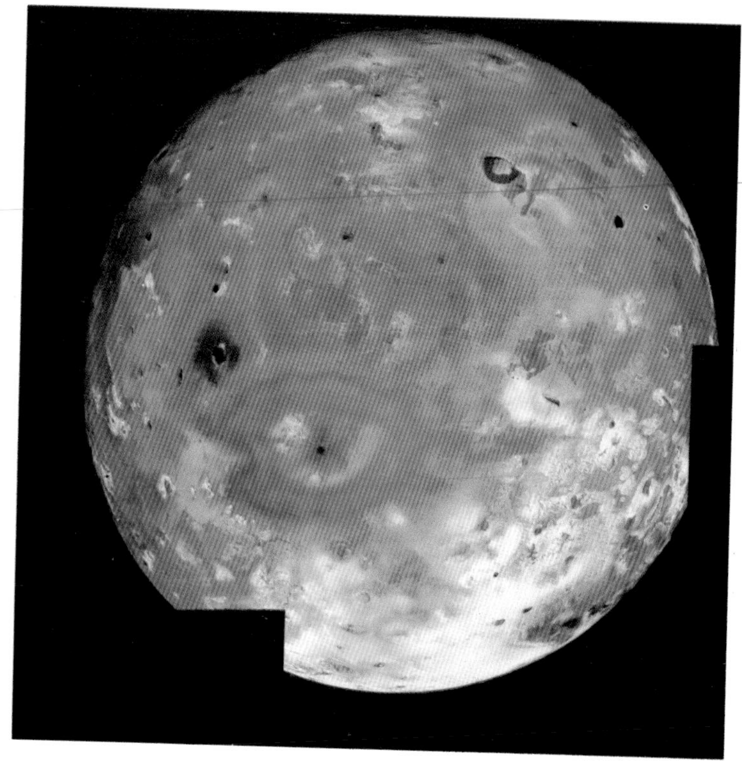

undertaken. It has demanded hardy and dependable spacecraft, precisely calculated trajectories, fail-safe communications systems, and a special kind of planetary patience.

It took the two spacecraft just short of two years to reach the vicinity of Jupiter, the closest of the outer planets, but the wait was quickly justified. Together, the two craft discovered a ring of solid debris that could have damaged or destroyed them but luckily did not. And in addition to the discovery of Io's volcano, they observed a previously unknown moon, bringing the total of Jupiter's known moons to fourteen. Yet it was the close-up images of the already familiar components of Jupiter and its satellite system that were the most dazzling and scientifically important.

Before Jupiter's gravity hurled Voyager 1 like a rock from a slingshot toward Saturn's largest moon, Titan—a year and a half away—it had produced 18,000 images of Jupiter, its ring, and three of the Galilean moons. Voyager 2 produced an equal number of photographs of the huge, landless planet; it rephotographed Ganymede, Callisto, and Io; and it took the first beautiful photographs of Europa, the fourth Galilean moon, covered by a miles-thick coat of ice, its surface etched by blood-red fractures resembling the capillaries near the surface of human skin.

LEFT: Callisto, only slightly smaller than Ganymede, is pockmarked by craters that are evidence of an ancient bombardment, perhaps the same one that scarred Mercury and the earth's moon. RIGHT: Io appears to be extremely volatile. Each of the small dark features in the photograph is a recently active volcano. The volcano with a bright halo near the center of the sphere was photographed in eruption by Voyager 1 a few hours prior to this photograph.

OPPOSITE: In November 1980, Voyager 1 photographed the intricate ring system of Saturn as the craft continued its journey toward the edge of the solar system and beyond. Visible in this photograph are shadows that the rings cast on the pale gaseous planet and on its moons Tethys and Dione.

Titan, Saturn's largest moon, as photographed by Voyager 1.

The decision, made long before either Voyager craft was launched, to send Voyager 1 to a near-rendezvous with Titan meant that, because of the resulting loss of an essential gravitational push from Saturn itself, Voyager 1 would not be able to visit either Uranus or Neptune. It would instead fly silently out of the solar system. But its sacrifice seemed very worthwhile. Voyager 1 relayed to earth thousands of images of Saturn and its intricate ring system before venturing to within 5,000 miles of Titan—the closest either Voyager craft would ever come to any planet or moon.

Titan was an enticing target. Before Voyager 1's arrival, it was known to be the only moon in the solar system with a substantial atmosphere. Voyager 1 was able to determine that Titan's atmosphere is a mixture of carbon, nitrogen, and hydrogen that is somewhat denser than the earth's atmosphere. It also revealed that, besides the earth, Titan seems to be the only planetary body yet discovered whose surface is at least partially covered with liquid. Titan's seas are not water, however. They are composed of extremely cold liquid methane that appears to evaporate and then condense in a rain of methane, much like the earth's water cycle.

Although Titan's size, solid surface, seas, and atmosphere superficially resemble the

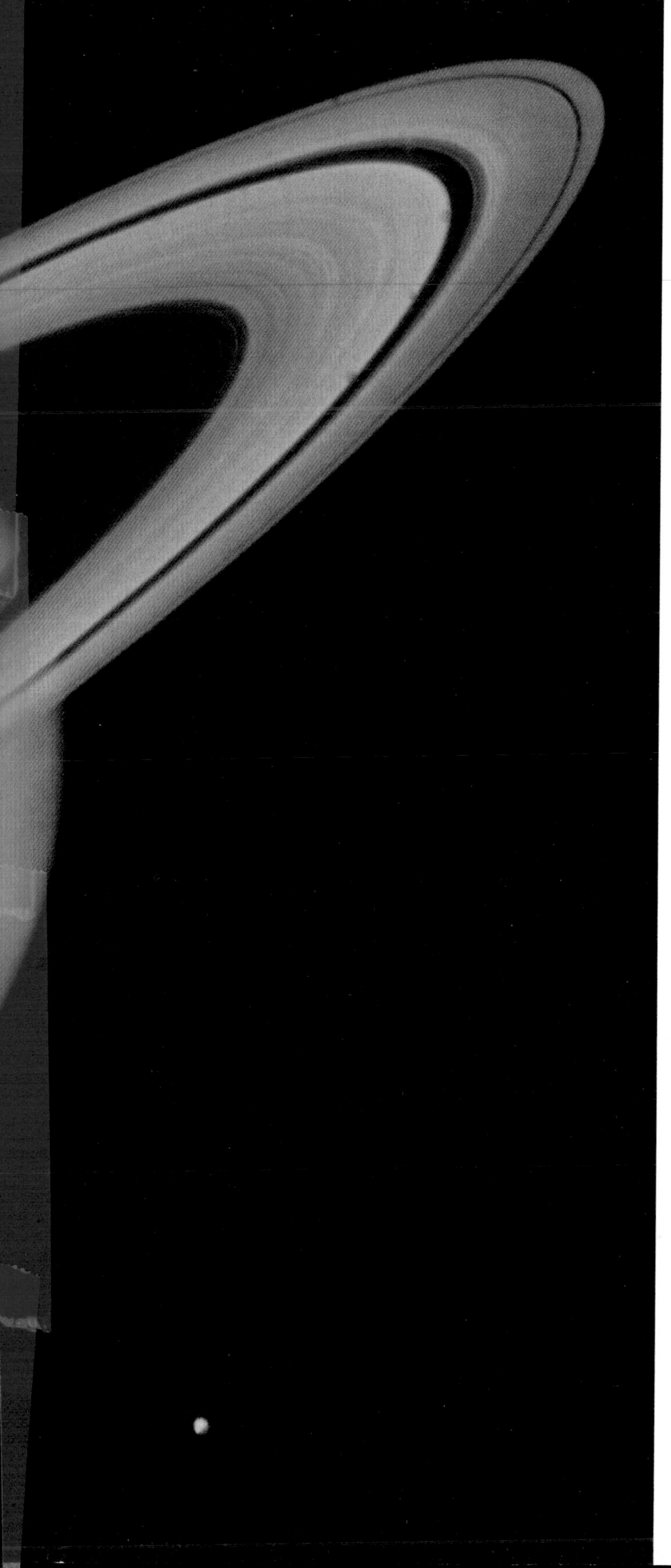

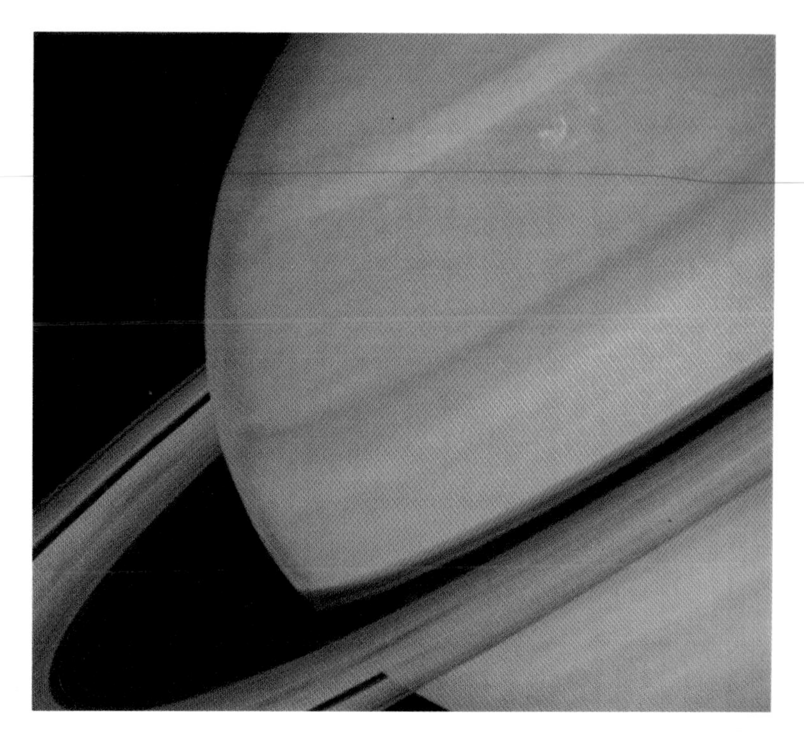

LEFT: To space travelers approaching Saturn, the planet would appear much as it does here. Saturn has much less intrinsic color than Jupiter, and only one quarter of the sunlight that bathes Jupiter reaches as far as Saturn.

ABOVE: In this color-enhanced photograph, made by Voyager 1, the blue band at Saturn's horizon is caused by the scattering of sunlight. The orange clouds are crystals of ammonia ice.

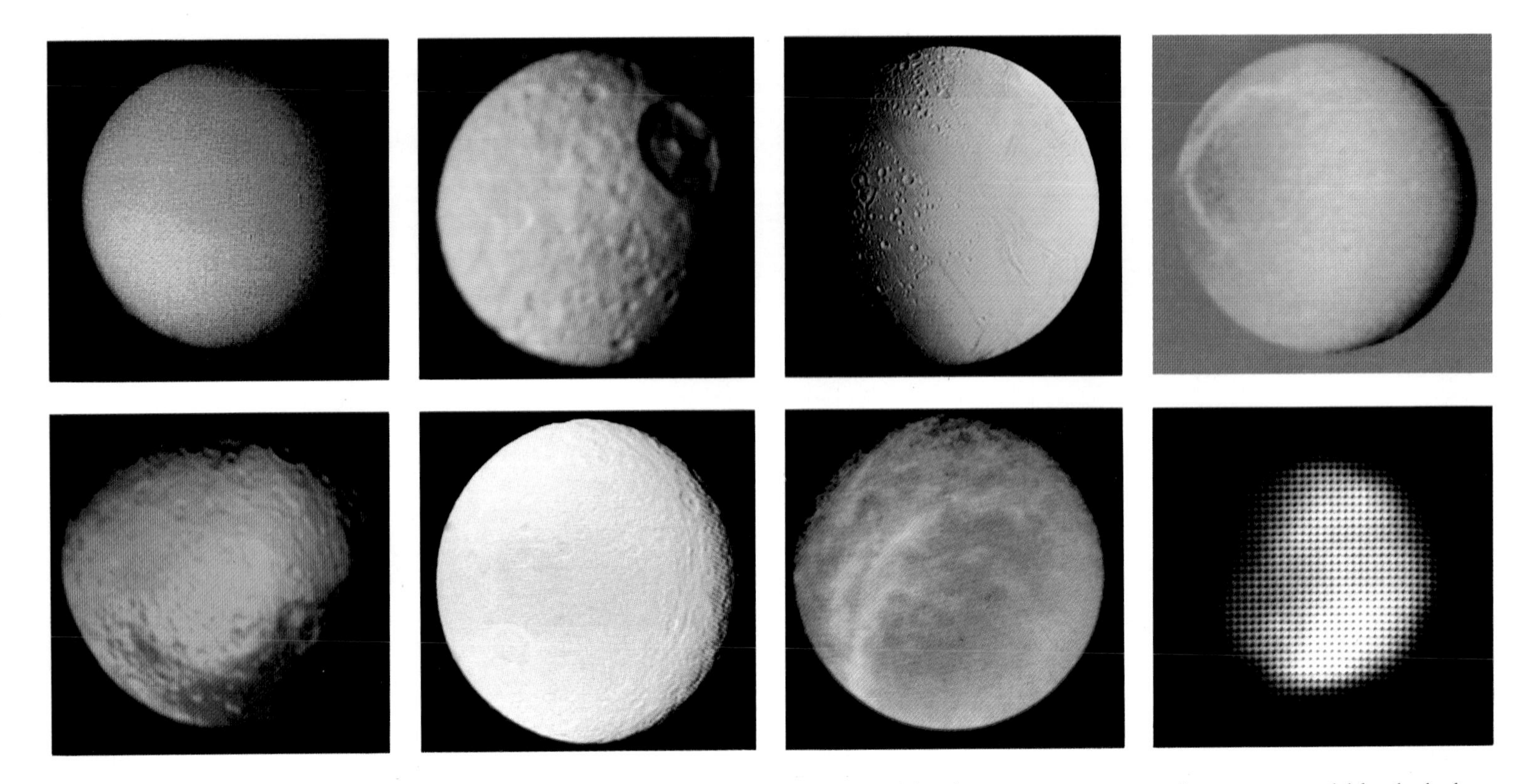

Voyager 1 and Voyager 2 photographed the seventeen moons of Saturn in such detail that astronomers can now surmise how these satellites evolved. Except for Titan—roughly the size of the earth and the only other planetary body yet discovered whose surface is partially covered with liquid—the Saturnian moons are all smaller than the earth's. TOP ROW, LEFT TO RIGHT: Titan, Mimas, Tethys, and Dione. BOTTOM ROW, LEFT TO RIGHT: Japetus, Enceladus, Rhea, and Phoebe.

earth's, it would doubtless be an inhospitable place to visit. Its methane air would be lethal to breathe, and its surface temperature appears to remain constant at about -500° F. And Titan's distance from the sun and its heavy atmosphere mean that its noonday surface visibility resembles the visibility on earth on a partially moonlit night. But this moon's current chemical composition may be able to tell us much about the formation of the solar system and particularly about the early evolution of the earth. Titan's atmosphere today, though it lacks oxygen, is considered by many scientists to be capable of producing the same organic molecules that are believed to have given rise to life on earth.

Following Voyager 1's fly-by of Titan and its spectacular visual investigation of Saturn's complex ring system in November 1980, the craft's trajectory sent it hurtling into the formidable darkness of the outer solar system. Voyager 1's scientific mission was complete, but it would be a traveler forever.

Voyager 2 is still at work. Its fly-by of Saturn followed Voyager 1's by nine months. Its support and observational systems operated to near-perfection while it was in the vicinity of Saturn, relaying to earth the discovery of three new moons—bringing the total of Saturn's known moons to seventeen—as well as thousands of important new images of the planet's intricately braided and banded ring system. Saturn's gravitational pull then flung the craft on toward Uranus, twice as far from the earth as Saturn is.

Voyager 2 will fly by Uranus in 1986, but it is now close enough to be able to confirm the speculation that Uranus also possesses a ring system; the craft has photographed at least nine distinct rings, and more may become visible as it nears the planet. In 1989, Voyager 2 will visit little-known Neptune—a planet so distant that the sun appears there only as a bright dot in the blackness. Then, like its predecessor, the craft will leave its own solar system, headed toward new suns, new realms.

Sometime in the twenty-first century, the two Voyager craft will sail past the limit of the sun's gravitational influence, and in the span of several hundred million years, they will silently circumnavigate the Milky Way, our own spinning galaxy that is merely a minor one among billions. It is very doubtful that either craft, its power long since exhausted, will ever be encountered by a civilization advanced enough to journey from its home to retrieve it. Yet because of that infinitesimal possibility, each Voyager starship was launched with a gold-plated copper phonograph record affixed to it, as well as a stylus, cartridge, and symbolic notation explaining how the phonograph can be used. The record and its jacket contain information about the earth's position, its present epoch, and the genetic composition of its human inhabitants. Included are greetings in sixty languages, the sounds of a humpback whale, and music from many cultures. Visual images are included as well, containing scenes of people creating tools and art, caring for children, and expressing emotion. This testament from the earth, our first statement to those who may inhabit other worlds, is expected to survive undamaged for a billion years. It will probably never be seen or deciphered, but that doesn't really matter. What is important is that traveling forever through the cosmos is a symbol of the intellect that has evolved on this planet. It is a symbol of a people, aware of the universe, who continue to explore.

THE THIRD PLANET

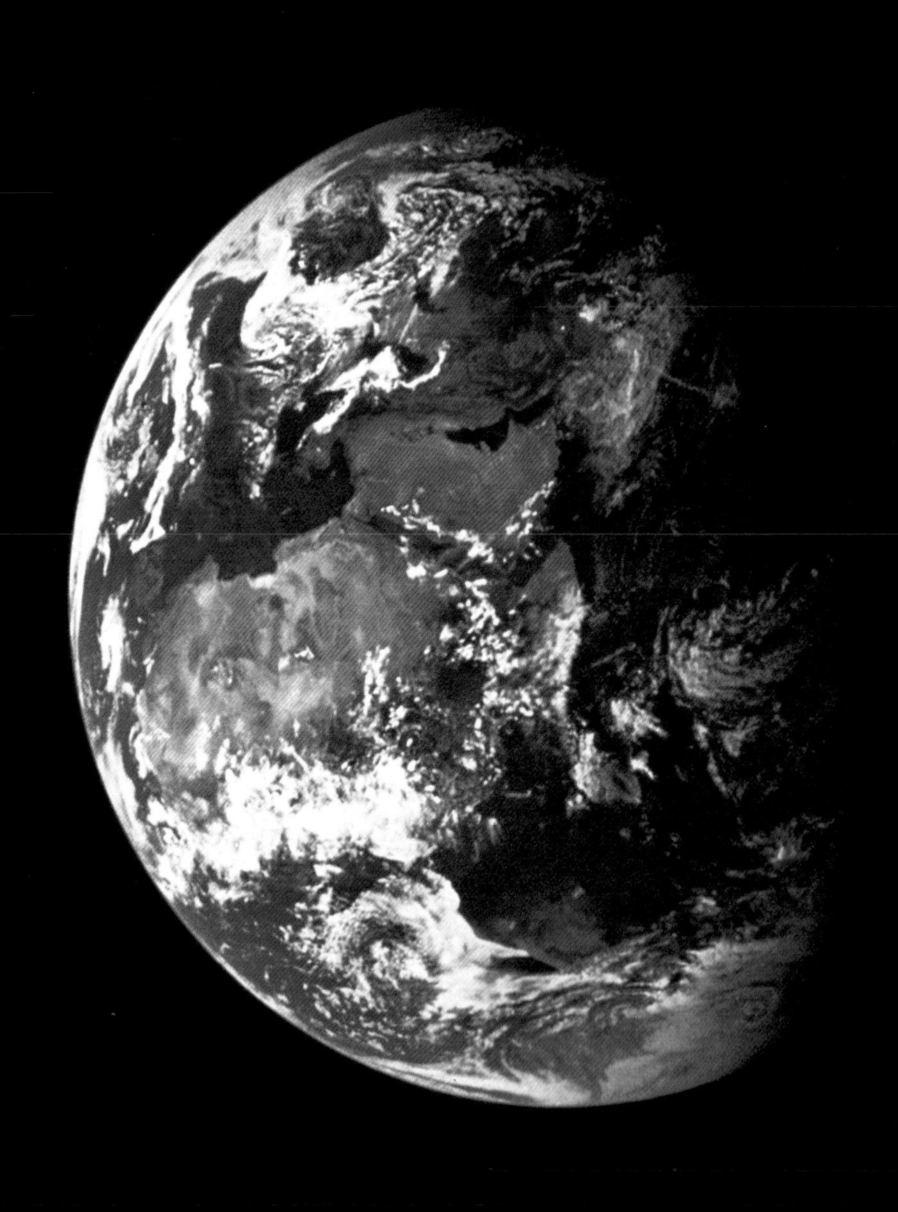

Viewing Earth from Orbit

The planet earth is the fifth largest in the solar system and third closest to the sun. It is a bright blue and white sphere that spins on its axis once every twenty-four hours. Its surface is covered predominantly by water; its land areas encompass verdant hills and valleys, glacial peaks, and wide, dune-swept deserts. The subsurface of the planet consists of active and unsettled solid tectonic plates that serve as underpinnings to its seven continents; its core is molten material, most likely iron. Its thin atmosphere, composed primarily of nitrogen and oxygen, carries enormous quantities of water vapor that circle the planet in swirling, solar-influenced cloud systems.

As inhabitants of this planet, we possess insights into its geology, its biology, and its place in the solar system. And we have understood for the last few centuries that the earth is a spinning sphere, that it is bounded by the black vacuum of space, and that it does not lie at the center of the universe—a place we now know to be unimaginably distant from us. Yet it wasn't until the first orbital flights of the early 1960s that anyone had ever had a direct view of the earth's curved surface. And it wasn't until the Christmas 1968 flight of Apollo 8, the first manned mission to lunar orbit, that three astronauts viewed personally—and the rest of the world saw in relayed television pictures—the earth as a round and bounded globe that is utterly alone in the emptiness.

The planet that astronauts Frank Borman, James Lovell, and William Anders observed from 260,000 miles away was starkly different from the cratered, monochromatic, and desolate moon. The spinning earth, seemingly suspended in a black lunar sky, looked like a lovely and delicate blue and white ball. There was no mistaking it—it was *home*. But there seemed to be a surprising fragility about it. It was small and beautiful, and it was the only object of shimmering, moving color within millions of miles. "On the way to the moon," Norman Cousins wrote about the Apollo flights, "man discovered the earth."

PRECEDING OVERLEAF: Clouds above the Pacific Ocean are lit by the sun in this photograph, taken by the author on the fifth shuttle flight.

OPPOSITE: Much of Africa is visible in this three-quarter view of the earth taken from Apollo 11 as it speeds toward the first lunar landing.

PAGE 160: An Agena docking target tethered to the Gemini 12 capsule is photographed as it crosses the coast of California. This view looks south toward Baja California and the Sea of Cortez.

PAGE 161: Skylab's S-band antenna is visible as the orbiting laboratory crosses the same point on the California coast.

The orbiter *Columbia* crosses over the Peron Peninsula and Shark Bay on the western coast of Australia on the fifth shuttle mission. The Indian Ocean is visible at bottom.

In an important sense, one of the most rewarding aspects of the first quarter-century of space flight has been this visual discovery of the earth—the realization of people everywhere that this really is a planet, a world entirely dependent on the sun, a world limited to its own vital resources. And since the Space Transportation System was initiated in 1981, the men and women who orbit the earth have been able not only to *see* the earth but to savor its awesome vistas. The Mercury, Gemini, and Apollo vehicles contained windows, of course—small and oddly shaped viewing ports through which astronauts could catch quick glimpses of the earth or stars when the capsule moved into proper position. Although much larger than its predecessors, Skylab had only a single viewing window in its living and working space, and its busy missions were focused inside the orbiting lab and not on the views outside. Each shuttle orbiter, in contrast, contains ten windows, each of which is twice as large as a passenger's window on an airliner, located in the forward cockpit, aft crew station, and overhead ceiling of the flight deck. In addition, an eleventh window, the only one that resembles a porthole, is located in the mid-deck external hatch. At least one of the windows on either deck offers an unobstructed view of the spinning earth almost all the time; only when the orbiter's black-tiled belly directly faces toward the planet is it obscured from view. In the midst of their work, as well as in idle moments, shuttle crew members are continually confronted by the earth. It is always the same earth that they watch—its seas and sailing clouds, its familiar land forms—but the vantage point from which they watch, through large windows 200 miles above the earth's surface, is startlingly new and endlessly fascinating.

As the shuttle vehicle orbits the earth, its velocity is 17,500 miles an hour, *thirty times* faster than the top speed of a commercial jetliner. The orbiter passes over the earth at 5 miles a second; in the time it takes for three deep breaths, the orbiter travels almost 100 miles. It circumnavigates the earth, a distance of some 25,000 miles, every ninety minutes. Yet this racing journey is smooth and utterly silent. There is no roaring wind to heighten the sense of speed, no passing blur of telephone poles. The only indication of movement of any kind is the visual relationship between the orbiter and the earth: To a crew member, the orbiter can seem motionless at one minute, hovering near a whirling earth; and moments later, to the same astronaut, the earth can seem as still as it does from an airplane, its quickly changing land and cloud formations evidence of the orbiter's amazing speed. But whichever perception the astronaut holds at any moment, there is a constant and pervasive sense of quiet, exquisite travel.

In this 1973 Skylab photograph, towering thunder clouds cast long shadows on the earth at sunset.

During its ninety-minute journey around the world, the orbiter passes from day to night, then back to day again. In the span of twenty-four hours, it travels through sixteen sunrises and sixteen sunsets, one every forty-five minutes—each one spellbinding, beautiful, and sudden. When the orbiter passes over the half of the earth that is hidden from the sun, it encounters a night that is deep and velvety black. Then, as it approaches the "terminator," the line on the earth that separates day and night, the first hint of light outlines the curve of the earth's horizon. In the course of only seconds, the faint light swells to bands of brilliant blue and red, formed by the atmosphere, bands that appear thick at the point where the sun is about to emerge and taper to thin lines at both extremes of the horizon. As the light suddenly intensifies, several distinct bands of dazzling color become momentarily visible. Then the bright fireball of the sun flashes above the horizon. Its fierce rays, unobscured by the atmosphere, bathe the orbiter in hot, glaring white light, yet the earth directly below the orbiter remains in the last moments of predawn darkness. Forty-five minutes later and halfway around the earth, the sun seems to swing to the horizon again and the beautiful bands of light reappear for a stunning instant, then vanish just as quickly, swallowed by the advancing darkness.

Each sunrise or sunset that the crew observes from the orbiter is boldly dramatic not only because it is so beautiful, but also because it is so sudden. Since the orbiter is traveling seventeen times faster than the earth revolves, a sunrise or sunset viewed from the orbiter occurs seventeen times more quickly than it does when viewed from the earth's surface. In much the same way that time-lapsed photography shows the growth of a flower as strangely fast and fluid, the orbiter's speed makes sunrises and sunsets appear so exquisite, so condensed in time, that crew members are repeatedly surprised by the seconds-long flurry of color and by the orbiter's dramatic transition from brilliant light to total blackness.

The blackness of each forty-five-minute night is absolute and enveloping. Depending on the phase of the moon and its position relative to the orbiter, night can be as utterly absent of light as the deep recesses of a subterranean cave. The pinpoint lights of the stars are clear, crisp, and steady—their atmosphere-induced twinkle is gone—but they are no brighter than they are on earth. They appear much as they do from the earth's high deserts, and they shed no discernible light on the pitch-black planet.

Seen at night from the orbiter's windows, the earth itself could easily be judged to be part of the black void of space. The darkness appears as though the earth itself had vanished, except for the occasional faintly glowing clusters of artificial light that mark its major cities.

The easiest way for an astronaut to find the earth in the darkness is to search for a disappearance of stars, to look for the curve of blackness seemingly cut out of the heavens. That blackness, that absence of even starlight, is the round and solid earth looming only 200 miles away.

There are instances, however, when the nocturnal earth makes its presence unmistakable. The pulsing red-orange light of major forest and grass fires stands out starkly against the blackness, for example, and electric flashes of lightning brilliantly illuminate from within the thunderclouds that generate them. The huge clouds seem to light up instantly and magnificently like enormous bulbs, and a single lightning bolt often seems to trigger a chain reaction of flashes from cloud to cloud so that the lightning appears to be walking its way for hundreds of miles across the darkened earth.

Lightning is a common sight during each short period of night, but during the forty-five minutes of day the sun's light on the earth is too intense for lightning to be visible from the orbiter's windows. Cloud systems, however, are vivid and three-dimensional. It is impossible for the crew to see through the layers of clouds to the earth's surface, but the billowing clouds themselves are fascinating. Tall, dark columns of thunderclouds seem to boil up through flatter layers and climb into the upper atmosphere; the leading edges of storm fronts seem to march forcefully across the ground. Clouds pushed by strong, high winds move into parallel streamers, like long white furrows, above the earth; thick clouds bank along coastlines; and above the open seas, huge storm clouds occasionally swirl into the fearsome circular patterns that can create violent hurricanes. Yet as volatile as the weather patterns are, individual storms and cloud systems remain easily identifiable as they move across the earth. Large weather systems and their characteristic clouds often remain intact for days, providing the crew members with dependable atmospheric "landmarks" that pinpoint their location over the earth.

There is seven times more ocean area than land area on the earth, and portions of the oceans are often heavily covered with clouds, so the crew sees only broad blankets of white clouds sheltering bright blue water. But when the clouds dissipate, the oceans show a surprising variety of color. The intensity of the blue corresponds to the water's depth—darkest where the sea is deepest, becoming a light blue-brown in the coastal shallows. The occasional dark patches of oil spills are easily visible, and the straight, white feathery lines that cut across the blue background are the wakes of giant ships.

OPPOSITE: Clouds blanket a watery planet, as seen from Apollo 11.

PAGES 168–169: In this photograph by the author, the sun brightly lights the tail and OMS pod of the orbiter *Columbia* as the earth below just begins to enter daylight.

PAGES 170–171: From orbit, the earth often appears to be formed of nothing but clouds and water, as shown in this photograph from the fifth shuttle mission.

PAGE 172: Streaks of clouds sweep across the Nile River and the Red Sea in this early Gemini photograph. The Gulf of Suez is visible at lower left.

PAGE 173: The tip of the Arabian peninsula is visible at left as *Columbia* crosses the Red Sea in this photograph by the author. The Gulf of Aden is visible at upper left, and the land mass at right includes portions of the African nations of Ethiopia and Somalia.

PAGE 174: Lakes Tangra and Tagje on the Tibetan plateau are visible in this photograph from STS-3.

PAGE 175: The Indian Himalayas, photographed on the second shuttle mission.

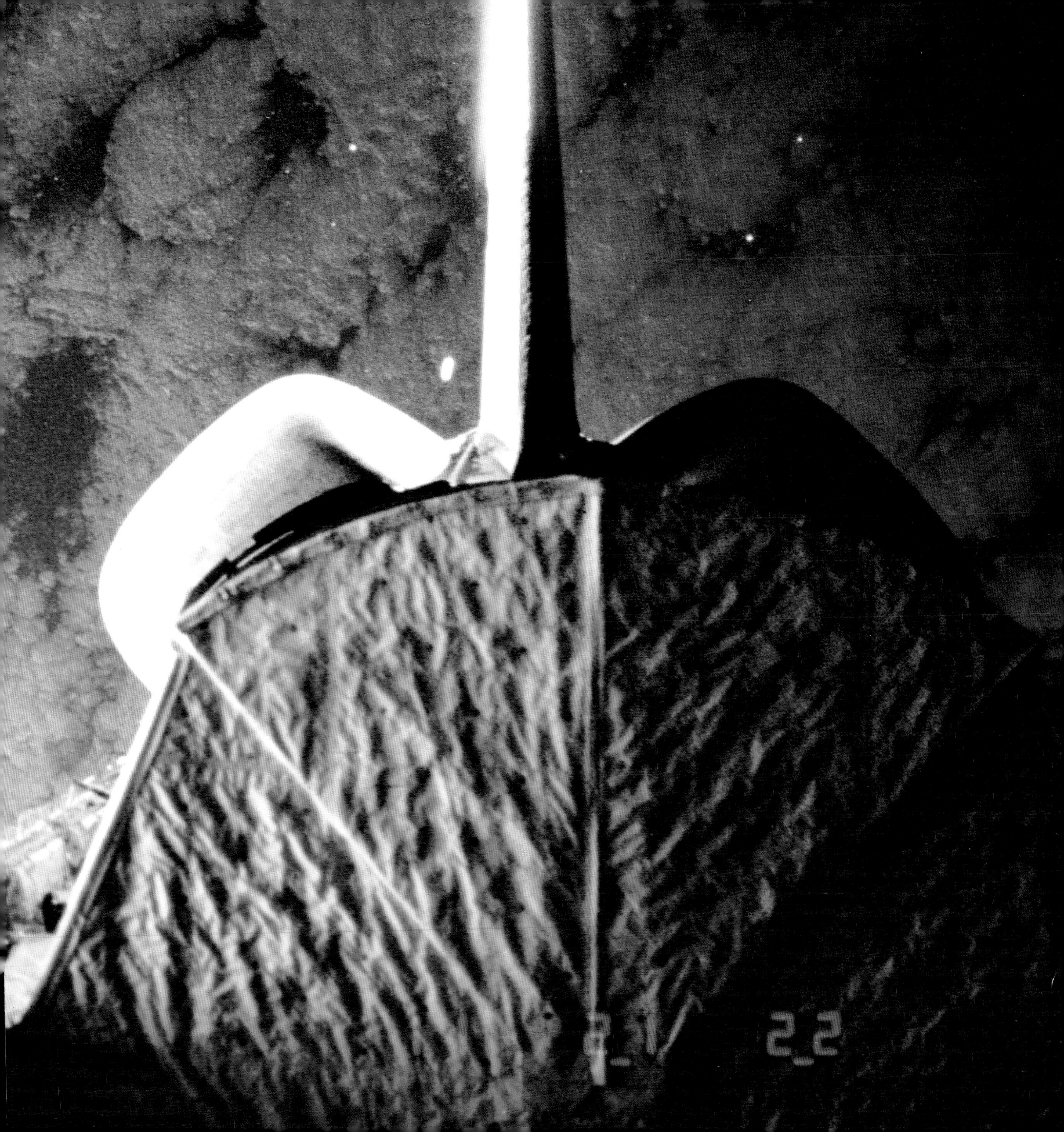

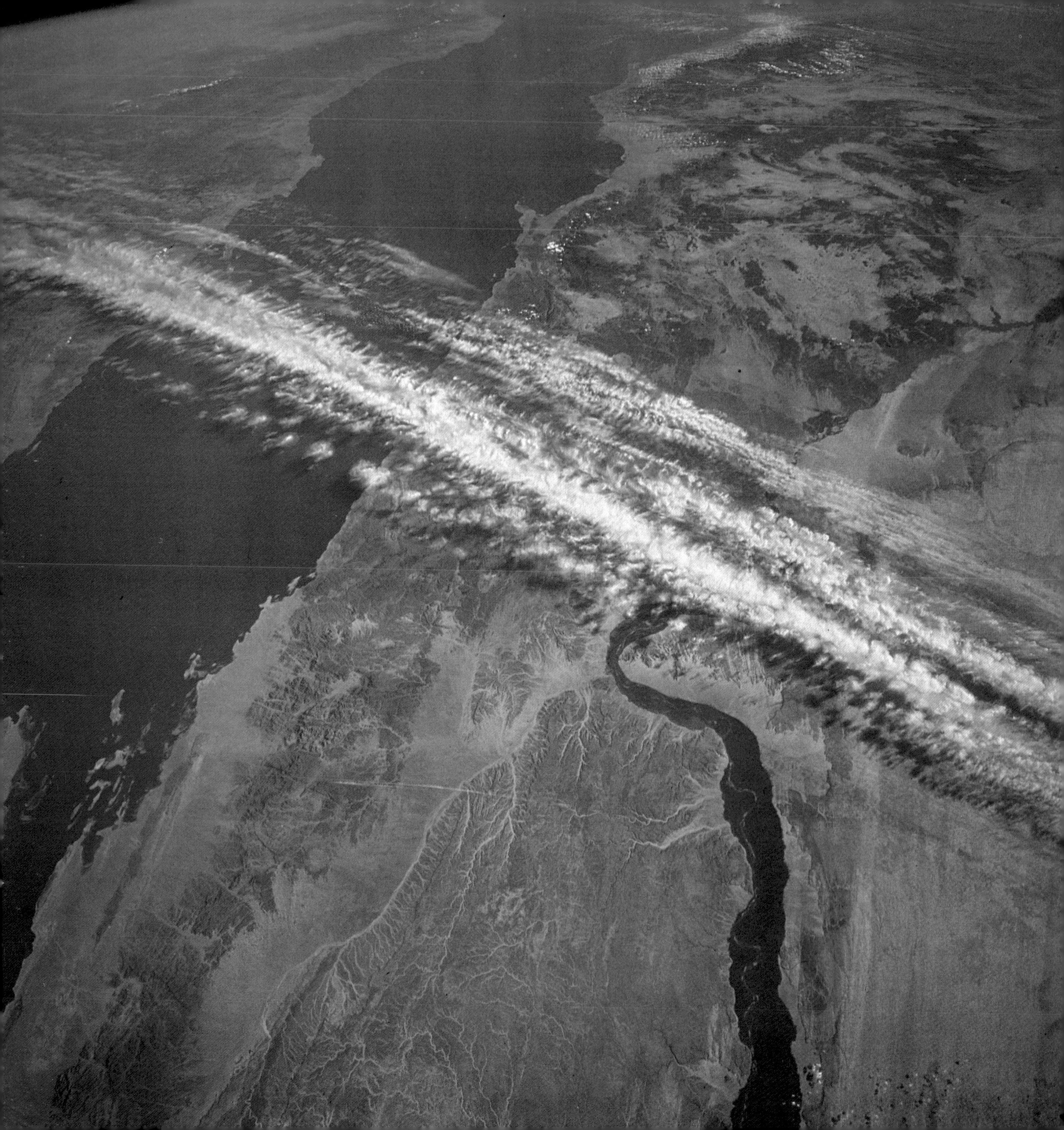

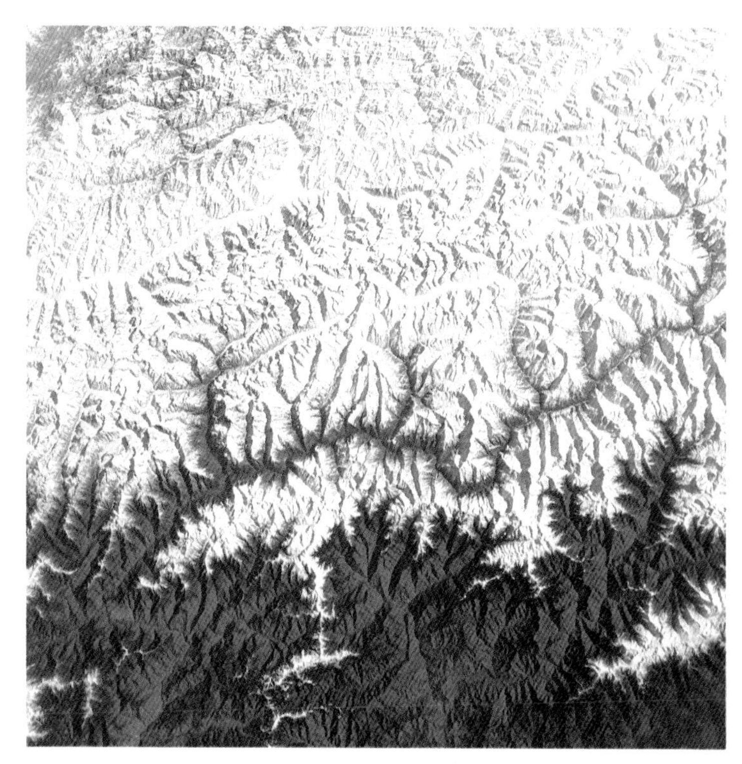

From an orbital altitude of about 200 miles, the breadth of the earth that is visible at any moment stretches almost 1,000 miles in a smooth curve from one horizon to another. As the orbiter begins to cross North America, for example, the crew can at once see from the Pacific coast across Baja California and the Gulf of California and into the Mexican interior. A few minutes later, a reach of the coast of the Gulf of Mexico from Houston to New Orleans comes into view; and soon virtually all of Florida is visible—from the gulf to the Atlantic, from the Keys to the Georgia coast.

Crew members don't have to look at the curve of the horizon to find proof that the earth is round. Even when they look straight down, they see not a flat, maplike piece of earth, but rather a gently rounded terrain whose curvature is made clear by the height of the orbit. This is a truth that everyone has known since early school days, but the visual evidence of it for the astronauts is nonetheless arresting: The waters of the wide Atlantic lie on a *round* surface; Australia *curves* from Perth to Sydney; the "flat" Midwestern prairies aren't flat at all.

Familiar land forms come into view as continual reassurances that the planet below is home—the long finger of Florida, Gibraltar, and Italy's boot, the thin line of the Malay

The snow-shrouded Himalayan range on the Nepali-Tibetan border as photographed on three shuttle missions: OPPOSITE, STS-1; ABOVE LEFT, STS-5; ABOVE RIGHT, STS-2.

PAGE 178: The Japanese islands of Honshu, Kyushu, and Shikoku as seen from *Columbia* on the second shuttle mission.

PAGE 179: The Society Islands, surrounded by the South Pacific.

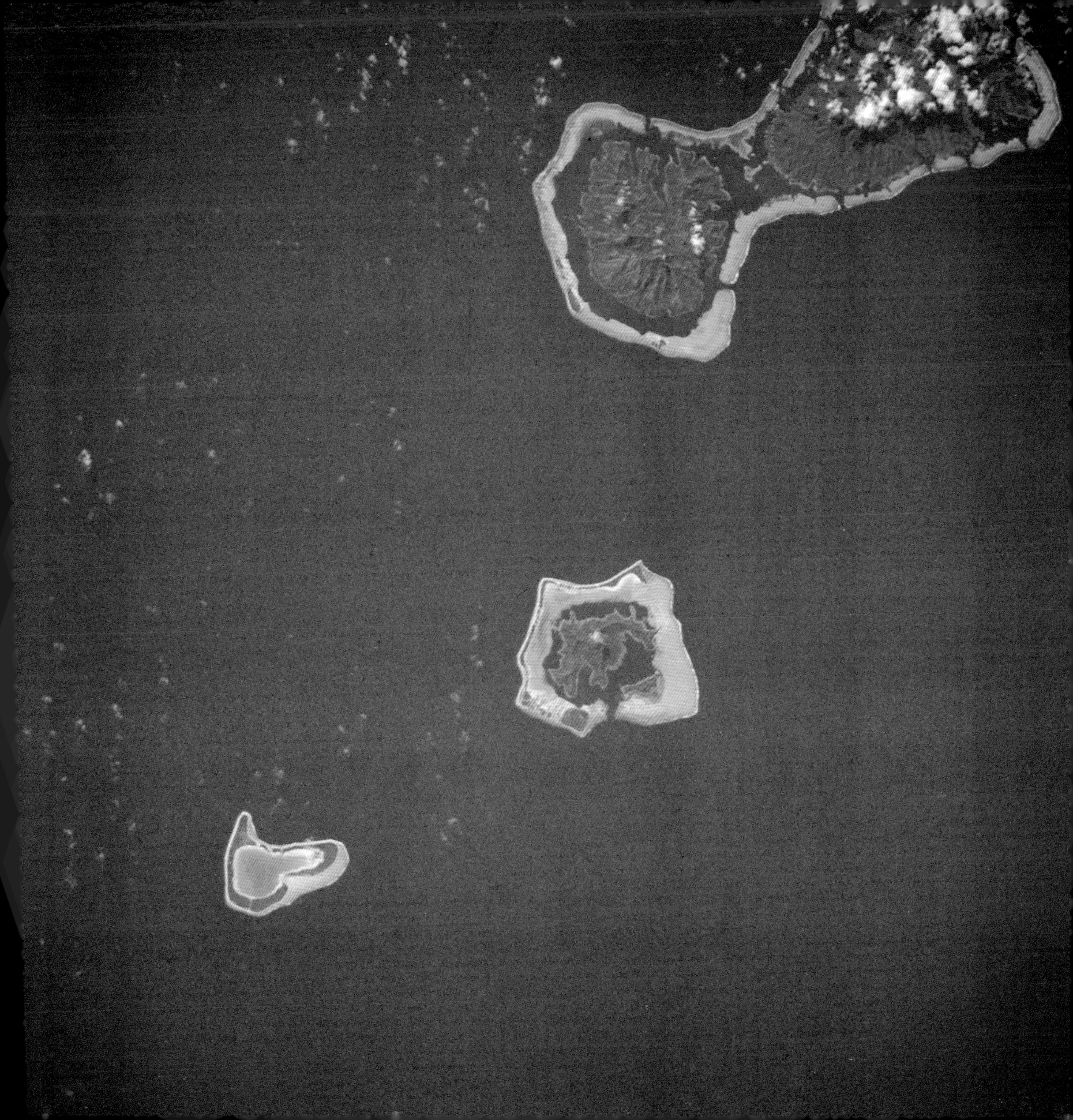

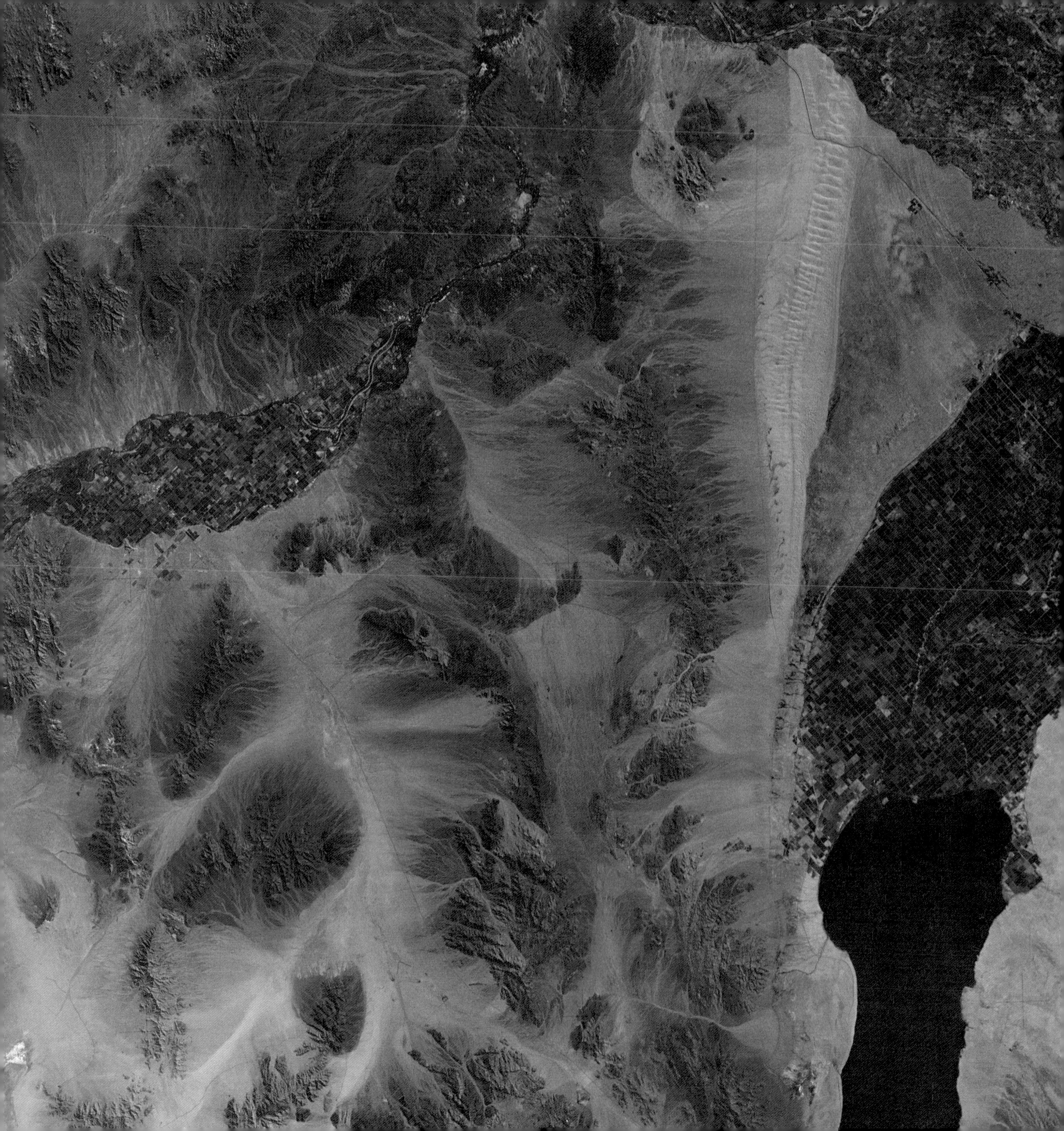

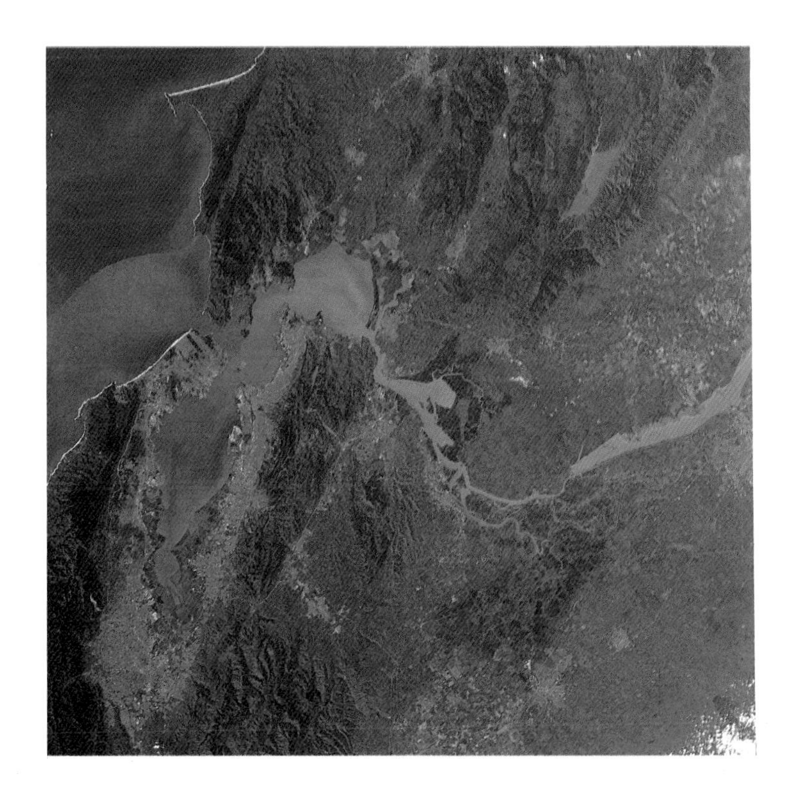

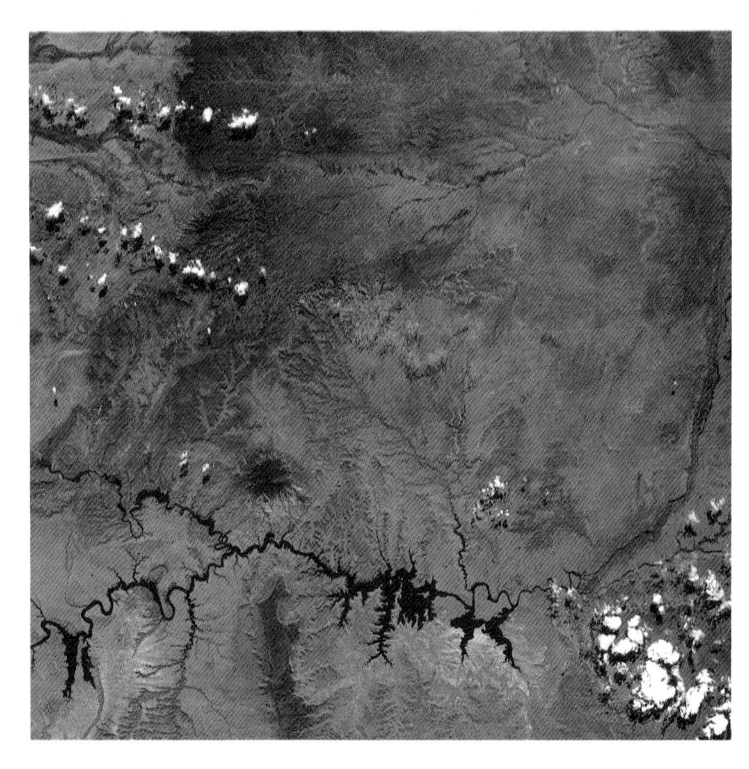

Peninsula, the emerald islands of Hawaii. But above much of the earth's land mass, especially above inland areas where distinctive coastlines cannot help with identification, it is difficult for an astronaut to ascertain exactly what is below by glancing down. The brown dunes of a desert could lie in America, Africa, or Australia; so could tawny grasslands. There are no easy gauges of direction—the simple distinctions between north and south disappear—except that on most flights the orbiter is traveling eastward, and its movement becomes a dependable compass point. As is the case inside the orbiter, however, there is no true up or down in the relationship between the orbiter and the earth. Depending on the orbiter's attitude (the direction its nose points), it can as easily seem to be below the earth—or beside it—as above it, and what *seems* to be is simply a matter of the astronaut's own perception.

OPPOSITE: The Chocolate Mountains and California's Coachella Valley are photographed from orbit. The Salton Sea is visible at right; the town of Blythe is near the upper edge of the photograph.

ABOVE LEFT: The San Francisco Bay area is photographed from Skylab.

ABOVE RIGHT: Lake Powell and the Colorado River in the high deserts of Utah and Arizona are photographed from Skylab in 1973.

This beautiful planet is home to nearly 5 billion people, yet despite the enormous populations that strain resources on nearly every continent, the crew members rarely see evidence of earth's inhabitants. Were they explorers from another world, they might even speculate initially that this watery world was uninhabited. No buildings are visible, nor are boats or cars, nor are humans or animals.

OPPOSITE: Chicago and the southern tip of Lake Michigan are photographed from Skylab.

LEFT and PAGES 184 and 185: Clouds, as seen from orbit, form sensuous patterns above the earth's desert regions.

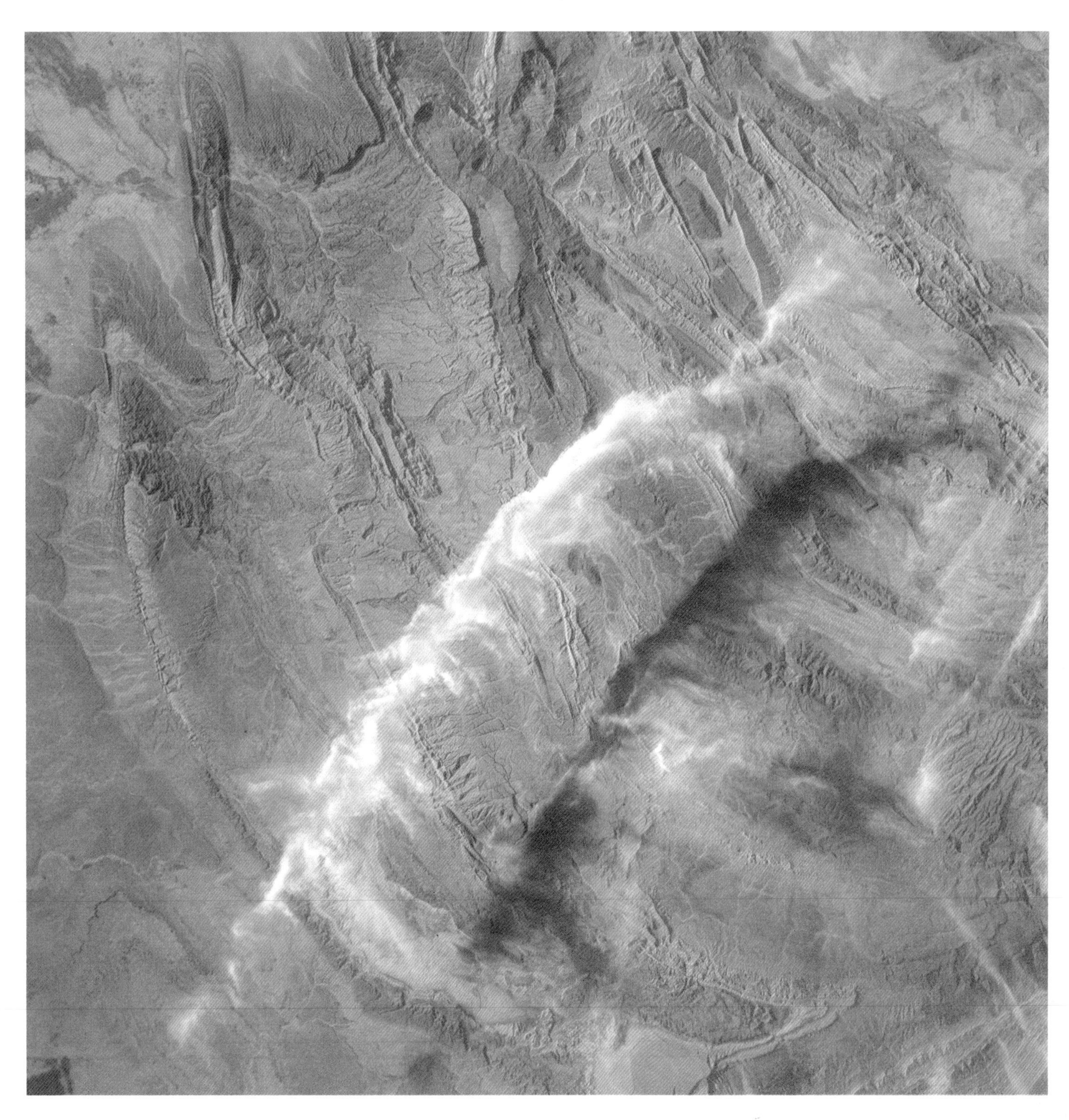

Apollo astronauts en route to the moon photograph the crescent earth against the solemn blackness of space.

It is humankind's penchant for creating straight lines that finally gives away the secret of habitation. The gridwork of cultivated fields, arrow-straight airport runways and taxiways, the uninterrupted lines of interstate highways, and the dimly visible, concentrated hatchwork of city streets—all lie in sharp contrast to the earth's own random, natural shapes. And there are other telltale signs: Thick smoke trails away from brush fires set by tribal peoples in Africa; bright green circular fields, irrigated by center-pivot sprinkler systems, contrast sharply to the buff Arabian desert that surrounds them; and the brief white threads of jet contrails float above the dark background of the land.

But it is the handiwork of the construction engineers that is the most vivid evidence of humankind's impact on earth. On flights whose orbital paths pass above Houston, crews can see the interstate loop that encircles their home city's perimeter and the highways that reach into downtown like the twisted spokes of a wheel. Looking closely, they can spot the runways at Intercontinental and Hobby airports and at Ellington Air Force Base; they can see Interstate 10 running east toward New Orleans and seemingly forever west across Texas, and Interstate 45 angling to the southeast, skirting the Johnson Space Center en route to Galveston Island; and they can quickly guess where their own homes must lie before the orbiter races out over the gulf.

The geography of Houston is familiar, somehow friendly from 200 miles up, yet the rest of the world seems hospitable as well. It is a breathtakingly beautiful sphere that from orbit shows no racial divisions, no political boundaries, no evidence of sorrow or strife. The earth is a serene planet from the vantage point of space, and the astronauts on each mission are privileged to be able, from time to time, simply to drink it in. It is the same earth they see and marvel at on foot, in an airplane, or from a boat on the open ocean—there is nothing new to see from orbit. Yet the *perspective* is so new, so inspiring, so humbling, that viewing the earth from the cold vacuum that surrounds it is inevitably one of the most remarkable and fulfilling experiences of every astronaut's sojourn in space.

Columbia
United

LEAVING SPACE

The Descent to Earth

The shape of the shuttle orbiter has been of no importance during its days in space. Were the orbiter the shape and size of the Empire State Building, it would travel just as fast through the vacuum that surrounds the planet. Once the orbiter was lifted beyond the earth's atmosphere, its orbital path would remain just as true during its many revolutions around the earth if it resembled a bicycle or a Buick. It is only now, as the orbiter is ready to return home, that its shape becomes critically and absolutely significant. During the thirty minutes between the time the orbiter first skims the upper reaches of the atmosphere and the moment it touches down on the long runway, its wide delta-shaped wings and its insulating exterior surfaces are essential to its safe return. And it is the orbiter's wings that make it a bona fide *aerospace* vehicle—one that is capable of maneuvering like an airplane as well as a spaceship, one that can fly in the earth's air as well as coast in the emptiness beyond it.

PRECEDING OVERLEAF: The orbiter *Columbia,* escorted by a T-38 training aircraft, crosses over California's Tehachapi Mountains en route to a dawn landing at Edwards Air Force Base at the end of the fifth shuttle mission.

OPPOSITE: The author photographs Bob Overmyer, pilot of STS-5, as *Columbia* is prepared for its re-entry into the earth's atmosphere.

The crew members prepare to leave space several hours before the deorbit procedures begin. They stow gear and secure instruments, close the cargo-bay doors, and retrieve the seats that have been folded up and taped into remote corners of the mid-deck since the vehicle first entered orbit. They float the seats back to their attach points, unfold them, and refasten the legs to the recessed fittings on the floor. Each crew member again dresses in a one-piece flight suit, boots, and helmet; each also wears a trouser undergarment called a G-suit, designed to prevent tunnel vision or fainting during the increasing G-force of re-entry. Each crew member also begins to drink up to a quart of slightly saline liquid—lemonade, grape drink, or salt water—as another precautionary measure against the faintness that could be brought on by the slight dehydration that has occurred during the days in zero gravity.

The commander and pilot, once again seated in the flight-deck cockpit, verify that the correct data are included in the deorbit and landing programs of the computers twenty minutes before the engines are scheduled to be fired. Following confirmation from Mission

Control that the craft is "go for deorbit burn," the commander maneuvers the craft to its deorbit-burn attitude. The orbiter responds by moving in a combination of pitch, roll, and yaw rotations that position its tail forward, aiming its OMS engines into the direction of flight.

As the craft passes over the eastern coast of Africa and heads across the Indian Ocean—halfway around the world from the planned landing site at Edwards Air Force Base in

California's Mojave Desert—the commander enters the deorbit-burn command into the computers. Fifteen seconds later, about one hour before the orbiter is scheduled to touch down, the OMS engines ignite and continue to burn for nearly three minutes; the precise length of the burn is determined by the mass of the orbiter and its payloads. If this mission had been scheduled to land at Kennedy Space Center in Florida, the burn would have begun 3,000 miles farther east, above a point halfway around the globe from Cape Canaveral.

Flight controllers in the Mission Control Room at the Johnson Space Center in Houston monitor the re-entry of a shuttle orbiter. Each controller monitors information telemetered to the ground from the spacecraft and is an expert in a specific aspect of the orbiter's operation.

Commander Bob Crippen, seated in the left seat in the orbiter's cockpit, monitors guidance- and navigation-systems displays during the fiery minutes of reentry. The bright glow outside the orbiter's windows is caused by the pounding of the speeding craft against molecules of the earth's atmosphere.

It is night when the burn begins. The astronauts, whose seats face forward, cannot see the bright, momentary tongues of exhaust shoot from the OMS engines, but they feel the onset of the burn and its unmistakable deceleration as they are pushed slightly back against their seats. It is a nudge of no more than .1G, gentle and steady, but it signals an inevitable return to earth. By the time the deorbit burn is complete, the fuel in the shuttle's main maneuvering system will have been consumed, making it impossible for the orbiter to regain its speed and stay in orbit. The burn slows the vehicle by only 200 miles an hour, little more than 1 percent of its orbital velocity. Even when the burn is complete, the orbiter is still traveling more than 17,000 miles an hour, but it is now in a newly shaped elliptical path—aimed, ready or not, for a fiery encounter with the atmosphere that will begin in less than thirty minutes.

After they confirm that the OMS engines are extinguished, the commander and pilot change the orbiter's attitude again, this time pointing its nose forward and pitching it up to a 40° angle of attack (the angle between the direction the orbiter's nose points and the direction it is flying). This angle will aim the broad underbelly of the orbiter, covered with thousands of superinsulating tiles, at the frictional brunt of the air.

In the minutes before atmospheric re-entry, the commander and pilot activate the auxiliary power units that drive the hydraulic systems that will, in turn, operate the orbiter's aerodynamic controls. Then they dump the remaining propellants from the forward reaction control system to help balance the orbiter for re-entry. Just before the atmospheric deceleration begins, some crew members inflate the legs of their G-suits. Bladders inside the suits squeeze their thighs and calves so tightly that they feel like they are being wrapped by boa constrictors from the waist down—tightly enough to prevent blood from flowing away from their brains and pooling in their legs during deceleration, which could cause tunnel vision or even blackouts.

Following the new orbital path determined by the deorbit burn, the shuttle coasts downward from its preburn altitude of 160 miles to a height of about 75 miles above the mid-Pacific, its speed increasing as it descends. Traveling close to twenty-five times the speed of sound—faster than any other type of winged craft has ever flown—the orbiter begins to graze the uppermost molecules of the atmosphere, pushing them aside with enormous speed and literally smashing them into atomic particles. As the wall of air grows steadily thicker, and as its friction against the orbiter increases, the shuttle's kinetic energy (the energy of its motion) is transformed into heat—so much heat that after only a few minutes of atmospheric contact, the temperature of the insulation covering the nose and the leading edges of the wings has reached 2500°F. This insulation and the thousands of tiles across the long, black, deltoid underbelly dissipate the heat as it is generated. But at the same time, the tiles—each one a thick silica-fiber composition coated with borosilicate glass—and the very hard non-conductive compound called carbon-carbon that covers the nose and wing edges shield the orbiter's aluminum frame and its passengers from the heat. The crew members, seated only a few feet from the blazing exterior, feel no increase at all in the cabin's temperature. They are riding safely inside a flying furnace, surrounded by the intense heat that completely destroys the vast majority of the thousands of meteors that encounter the atmosphere each year. But the shuttle orbiter is a meteorite that survives this conflagration virtually undamaged.

For the three crew members seated on the flight deck—the commander, pilot, and, positioned behind them, a mission specialist who serves as a flight engineer—the first real indicator of the orbiter's re-entrance into the atmosphere is the quivering needle of the G-

OVERLEAF: The author photographs commander Vance Brand, left, pilot Bob Overmyer, and the fiery glow of re-entry as *Columbia,* traveling at about Mach 15, makes its meteoric descent over the Pacific Ocean en route to a California landing.

meter. For days, this needle has been fixed at zero, as if it were painted on the dial. Now it shudders to life and slowly begins to rise. Then there is an unmistakable whisper of rushing air, at first almost too faint to hear, then louder and louder still. A faint red glow appears at the edges of the cockpit windows, then spreads across them and seems to curl up over the fuselage. Steadily, relentlessly, the increasing pressure of the astronauts' bodies against their seats, the noise, and the glow crescendo.

The perpetual floating that had become nearly second nature during the course of the mission has now slipped away irretrievably. The astronauts are no longer floating in their seats; they are sitting down hard against them. They feel increasing pressure on their thighs, buttocks, and backs as the deceleration continues. Beads of sweat on their foreheads run *down* into their eyes, and they grow increasingly aware of their bodies as the deceleration reaches the maximum force of 1.7Gs. The flight engineer, who weighs 150 pounds on earth and has weighed nothing at all for a week, now, because of the deceleration force, weighs more than 250 pounds, but he *feels* even heavier. None of the crew members is in real discomfort, but none of them finds the body-compressing force a pleasure.

The noise of the rushing air as the orbiter slows to a speed of less than 15,000 miles an hour increases to a hoarse growl and then to a roar. The glow outside the windows—produced by air molecules that are being buffeted so much that they emit light—grows steadily brighter, the red turning to orange, to rose, and then to a hot, pale, pinkish white; it is as if the orbiter were flying down an enormous, noisy neon tube. The hurtling orbiter slams against the air with such ferocity that it strips the electrons from the air molecules. This phenomenon creates an electromagnetic cone around the vehicle that blocks radio waves and interrupts communications with the ground, a blackout that lasts for about twelve minutes during the period of maximum heat and that ends as the orbiter slows to a speed of about 8,000 miles an hour, still 180,000 feet above the earth.

Prior to the advent of the shuttle era, every spacecraft that returned to the earth re-entered the atmosphere much like a cannonball would—plunging headlong, almost straight down, through the air. The friction of the air created incredible heat, and the craft braked very quickly. The bell-shaped Mercury, Gemini, and Apollo modules fell so steeply from orbit to splashdown that their crews had to endure forces equal to about 5Gs. Astronauts who were homeward bound from lunar missions could experience a punishing 8Gs as their modules shot back into the atmosphere at more than 24,000 miles an hour. And

OPPOSITE: *Columbia* is photographed by astronaut Kathy Sullivan from a T-38 training aircraft as the orbiter makes its final turn before touching down on the runway. Elevons that control the craft's flight through the air are visible at the aft edges of the wings. Below the tail rudder, the three main rocket engines used at liftoff are visible; above them, to the right, is one of two orbital maneuvering system rockets. Located between the wings and beneath the cones of the main engines is a movable body flap that is used to balance the orbiter's attitude during atmospheric flight.

PAGE 200: The three main parachutes safely slow an Apollo command module for its ocean splashdown as it returns from a mission to space.

PAGES 200–201: Safety and rescue divers prepare to open the hatch of an Apollo command module that has returned from a trip to the moon. The divers have placed a flotation collar around the girth of the module. The three large flotation balloons are deployed automatically as the module splashes down to ensure that it floats upright. (James L. Long Associates)

OPPOSITE: The orbiter *Challenger* touches down in darkness on a runway at Edwards Air Force Base. The flight of STS-8 in August 1983 included the first night launch and landing of the shuttle era.

Columbia lands at Edwards on completion of its fifth shuttle mission in November 1982, after orbiting the earth 81 times. Runway lights are visible in the foreground.

in every splashdown re-entry, the G-force of atmospheric deceleration was punctuated by a hard jolt as drogue parachutes were deployed, a harder jolt as the main parachutes opened, and a final hammerlike smack as the blunt heat shield of the module plunged into the water. The couches in which the astronauts lay were even fitted with shock absorbers designed to temper these sudden jolts.

By comparison, the trip home in the shuttle is an easy ride; there is a much gentler descent and far slower deceleration. Yet re-entry remains a very critical part of the flight. Spacecraft have been successfully rocketed into space for almost thirty years, but the era of returning them to the ground aerodynamically is still in its infancy. The orbiter's angle of attack must be positioned very precisely in order for the craft to survive the fiery slide through the upper atmosphere. And in addition, the 100-ton, unpowered glider must reduce its airspeed to a comfortable 200 miles an hour for landing, must descend to an altitude just inches above the ground, and must steer a path to a prearranged spot at the approach end of a specific runway. It performs these tasks by making a series of dramatic, high-banked, hypersonic S-turns as it sweeps homeward across the world.

As the sun sets, following a daytime landing, the orbiter, its mission complete, is surrounded by servicing equipment.

OPPOSITE: Following its fifth mission, *Columbia,* reflected in a puddle on the lake bed at Edwards Air Force Base, is towed toward the facility, where it will be mounted on a 747 jetliner and returned to the Kennedy Space Center.

As it approaches the coast of California, the orbiter is still traveling many times faster than the speed of sound. The autopilot, using instructions within the computers, steers the hurtling machine into a series of high banks, rolling it as much as 90° onto its side, a maneuver that drops the orbiter into the thicker, slower air below. It then reverses the bank by rolling to the other side, dissipating its speed much like a skier controls his or her descent down a slope by making a series of sweeping turns. As the orbiter banks into a left turn, the crew on the flight deck watches in the forward windows the distant, dim horizon turn until it is almost vertical. Then, as the autopilot commands a bank reversal from left to right, the crew watches the line of the horizon roll until it is momentarily level again, then continue to roll until the right limb of the horizon disappears out the top of the window and the left limb falls away and is hidden by the orbiter's nose.

The orbiter's reaction-control thrusters become less efficient and are gradually phased out as it descends into denser air. Its aerodynamic controls—the movable trailing edges of the wings, called elevons, that control pitch and roll, and the split rudder on the large vertical tail that controls yaw motion and acts as a speed brake—become increasingly effective. The

Columbia
NASA
United States

orbiter is now an airplane; its wings and tail steer it through the air. Its speed continues to drop dramatically and its angle of attack is being steadily reduced, but it is still falling much more like a rock than a graceful, gull-like glider. Some gliders can fly 40 feet horizontally for every foot of drop; a jet airliner flying unpowered would fly about 15 feet forward for every foot it dropped. The best the orbiter can manage is 4½ feet of horizontal flight for each foot it falls.

As the orbiter crosses Monterey Bay, the sun suddenly lights the eastern horizon and then floods the sky with pale light, erasing the last traces of red glow from the orbiter's windows. The ground-support personnel and spectators at Edwards hear two sharp sonic booms as the orbiter approaches from the west and crosses above the Tehachapi Mountains. And then suddenly they can see the orbiter—a bright dot almost directly overhead in the morning sky. For the astronauts inside the orbiter, the final 40,000 feet—the final seconds of their flight—pass with quick precision.

The orbiter flies above the runway, then arcs around the curve of a designated "heading alignment circle," 7 miles beyond the runway, that aligns the craft for its final descent. The commander takes control of the vehicle from the autopilot and familiarizes himself with the orbiter's responsiveness in the air (the orbiter can land automatically, but the astronauts and NASA officials still prefer to use manual control). With his hand on the hand controller in front of him—now nothing more than a pilot's "stick"—he aims the nose of the orbiter at a point on the ground a few thousand feet from the runway. He then uses data sent up by radio beam from the runway below—a microwave landing system that guides the orbiter toward the runway in much the same way that a commercial jet's navigation system operates. The commander monitors the vehicle's airspeed and altitude on a Plexiglas display screen that is positioned in the window so that he does not have to take his eyes away from his target. The pilot begins to call out a cadence of airspeed and altitude information.

Seconds before touchdown, as the orbiter slows to 300 miles an hour—about 1,700 feet above the ground and about half a mile from the runway—the commander lifts the nose of the orbiter to initiate the landing flare, and the steep approach path flattens to a more gentle angle. At 400 feet above the ground, the pilot lowers the landing gear.

Then the orbiter passes a few feet over the end of the runway, its rudder now open in a wide, braking V. The main landing gear are now ten feet above the surface; the orbiter is flying 230 miles an hour. Then "5 feet . . . 4 feet . . . 3 feet . . . 2 feet . . . 1 . . . touchdown." The main gear settle onto the runway, the tires spitting dust and smoke, the orbiter traveling

NASA officials await *Columbia*'s crew members, who will soon leave the vehicle that has carried them millions of miles in earth orbit.

Looking like a strange prehistoric bird, the orbiter is connected to servicing umbilicals that supply it with power, purge gases from its engine compartments, and cool it down after its desert landing.

215 miles an hour. As it slows and the air no longer supports its raised nose, the forward landing gear falls with a jarring *whump*. The wheel brakes are applied, and the orbiter, now no longer a flying machine, rattles and shakes and lumbers down the runway, rolling a mile and a half before it finally stops.

It is done. On their headsets, the astronauts hear the congratulations of the flight controllers in Houston; they acknowledge them and then go back to work, preparing to turn the vehicle over to the ground crews. Outside, the dust whipped up by the landing still hangs in the morning air, the light wind rustles the dry grass at the edge of the runway, and a black crow curls into the sky. But the orbiter, the machine that has traveled more than 2 million miles through the sweeping void of space on this mission, is now motionless, silent. A spaceship has landed on earth.

A spaceship has landed on earth. The orbiter sits motionless on Rogers Dry Lake bed in California's Mojave Desert. Its journey in space has ended.

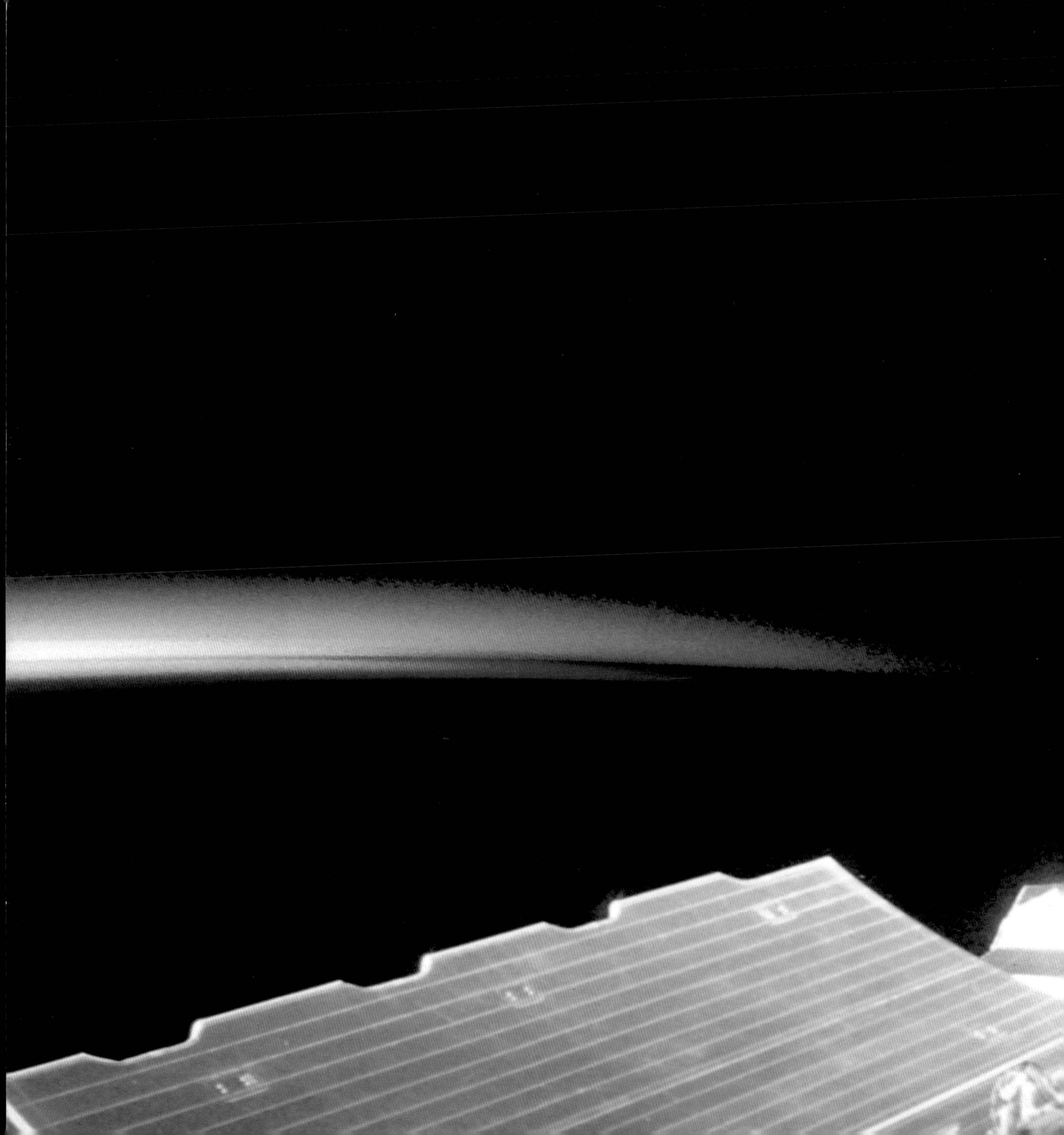

THE PROMISE OF SPACE

A Conclusion

PRECEDING OVERLEAF: The sun brilliantly lights a satellite sunshield and a radiator mounted on the gunwale of *Columbia*'s cargo bay in this dawn photograph by the author.

OPPOSITE: An abandoned launch pad at Florida's Kennedy Space Center, which was used for the launches of the final Mercury missions in the early 1960s, has already become a ruin of the space age. (Hans Teensma)

Along the thin ribbon of sandy beach at Cape Canaveral, rusted gantries rise like ramshackle steel skeletons. The abandoned launch complexes of the Mercury and Gemini programs, shrouded by thick green undergrowth, are battered by wind and water and salt. At Complex 5–6, from which Alan Shepard was rocketed into suborbital space in 1961, a gaunt gantry still stands. It is surrounded by the strange and archaic paraphernalia that made that first launch possible and by a collection of early Snark, Mace, Navaho, Redstone, Atlas, and Titan rockets—each one poised vertically for the appreciation of the more than 5 million tourists who visit the cape every year.

Visitors are forbidden to visit Complex 14, the site of the launch of John Glenn's first orbital flight and of the remaining Mercury missions, but there would be little for them to see if they did visit. The gantry is gone and the pad has been dismantled. Grass grows on the cement-and-sandbag bunker that once protected launch observers, and "ABANDON IN PLACE" signs hang on the leaning metal walls beside the pad. Marine paint that once safeguarded the structures from the coastal weather peels away in twisting strips, and the concrete ramp on which the launch vehicles were rolled into position is now buckled and cracked. Pale green moss clings to the seaward side of the ramp, and small flowers grow out of the crumbled and powdered concrete.

It is hard to imagine that Complex 14 is already a ruin of the space age, an obscure archeological site from little more than twenty years ago. And it seems ironic that this derelict location, a historical site that has been "abandoned in place," lies only three miles south of Complex 39, where the shuttle orbiters are continually refurbished, reloaded, and launched. Those three miles represent less than three decades of experimentation into the means of getting to space, of exploring that limitless new frontier, and of using its unique environment to enhance our lives and challenge our spirits. In less than thirty years we have gone from cautious uncertainty about whether people could even survive in space to confi-

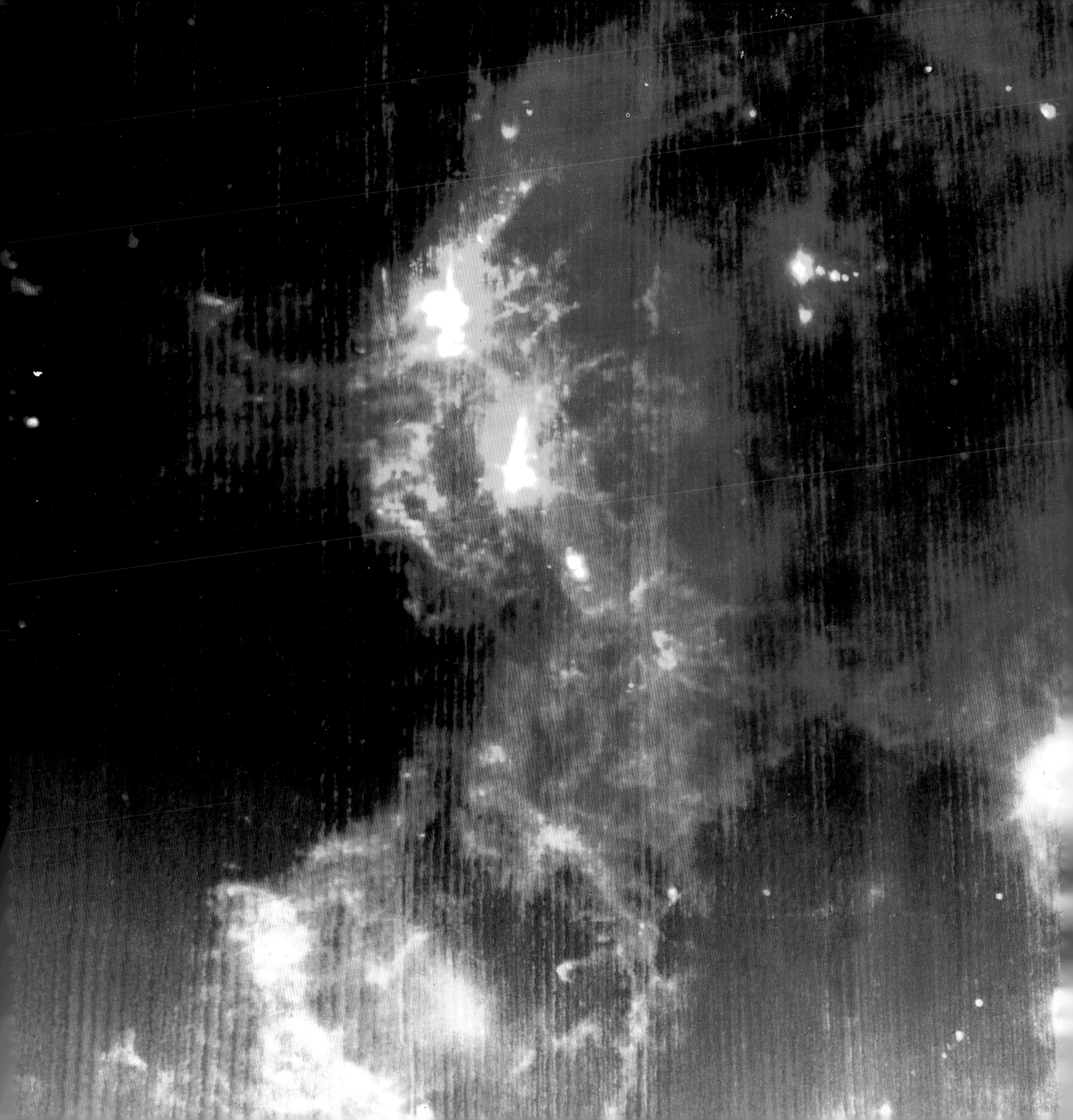

dent optimism that the human presence in space will soon be as commonplace as the presence of travelers on the seas and in the skies. Not even the most clairvoyant space scientists or science fiction writers can be sure what the shape and scope of our ventures into space will be thirty years from now, but it seems safe to speculate that we will be entering space with great regularity, that large numbers of people will be making the journeys, and that the term "astronaut" will have become a thing of the past.

During the pioneer era of aviation, only a small and select group of people flew in airplanes. For those too young to remember, books and films have preserved the dashing image of brave men and women in leather helmets, goggles, and long scarves who flew rickety biplanes to deliver supplies and mail to remote locations, to spray crops, and to barnstorm above the crowds at county fairs. Those first flyers were known as "aviators"—a term imbued with pride, strength, and daring. Today, the term "aviator" seems as archaic as the wooden-winged planes and the scarves that flapped aloft in the wind. The people who fly today are known as pilots and passengers; they are captains, first officers, and flight atten-

OPPOSITE: The IRAS satellite produced this falsecolor image of the constellation Orion. The differing intensities of infrared radiation are converted to reds, greens, and blues. The large, bright feature at the top of the photograph is a region of ionized hydrogen gas and dust heated by the star Bellatrix.

A Deep Space Antenna, one of three positioned around the world, communicates with the unmanned spacecraft that probe the solar system's outer planets, and listens for radio signals that emanate from the distant reaches of the universe. (James Balog)

dants, as well as bankers and ballplayers, lawyers and doctors, grandparents and technicians and tourists. There are no longer any aviators among us.

A similar shift in the terminology we use to describe the people who go to space is already under way. In the shuttle era, so many people go to space to do so many kinds of jobs that the term "astronaut" alone no longer adequately describes the individuals or their duties. The principal positions on each shuttle flight are, of course, commander, pilot, and one or more mission specialists; in addition, payload specialists fly on a wide variety of experimental missions. Yet it won't be long before pharmaceutical technicians, botanists, welders, journalists, and artists begin to ride the spaceships into orbit; and it will make no more sense to call them astronauts than it does to call someone who rides in the tourist class of a jetliner an aviator. By the turn of the twenty-first century, the term "astronaut" will conjure up images only of the pioneers of space, and will relate not to the exciting and unknown future, but to the proudly remembered past.

The near future of the Space Transportation System promises expanded operation and continued experimentation. The shuttle orbiters *Columbia, Challenger, Discovery,* and *Atlantis* are expected to fly a combined total of twenty to twenty-five missions a year by the end of the 1980s, meaning that a shuttle will be launched almost every two weeks and that some astronauts may fly as often as four times a year. Satellite deployment will remain a fundamental component of the shuttle program, and satellite retrieval and repair will become increasingly common. Spacelab missions will undertake a wide array of experimental projects, and the manufacture of alloys and pharmaceuticals in weightless conditions will have begun. The Space Telescope will be lifted into orbit and will begin its search into the cosmos, and ongoing earth-science investigations will continue to focus on the planet closest at hand. And inevitably, the shuttle will live up to its name when it begins to *shuttle* crews and supplies from earth to an orbiting space station, an outpost that will surely become our gateway to the solar system.

The first permanently occupied outpost in space will probably look like something a child would build with an erector set—an ungainly collection of scaffolding, solar collectors, and cylinder-shaped modules that will be ferried into orbit in the cargo bays of the orbiters, then assembled with the aid of free-flying tractorlike devices called tele-operators. The first experimental space station will provide living quarters for four to twelve people at a time, and individual crew members are likely to be stationed at the outpost for three-month periods. A command module, containing the crew's life-support equipment, will probably be connected to at least one unmanned module or platform containing astronomical and earth-science observational equipment and to a large manned laboratory similar to the current Spacelab. The modular design of the outpost will permit expansion of its size and the scope of its endeavors simply by the addition of new modules, platforms, antennas, or power-generating systems. One aerospace company has already developed an automated beam builder, small enough to be carried into orbit in the cargo bay of the orbiter, that can extrude ultralight triangular support girders from coiled metal plate. The beams, which on earth would weigh less than a hundredth the weight of standard, comparably sized terrestrial girders, would be essential structural elements of a large, multifaceted space station.

When this first space station is in orbit and operational, the hundreds of experiments and projects that it will make feasible will initially fall into four general categories: materials-processing research, akin to the studies in progress in Spacelab; observations of the earth's climatological, geological, agricultural, and other conditions; studies of the effects of weightless environments on all forms of life; and investigations into the intriguing potential applications of solar power. Sun-generated electricity will be essential to the operation of the space station, as it is now to the operation of hundreds of communications and observation satellites, and NASA planners are investigating whether large-scale orbiting generating plants may even be able to supply the earth's electrical needs. Some planners speculate that by early in the twenty-first century, electricity generated in geosynchronous orbit by enormous arrays of solar cells and transmitted to earth could meet most of the United States's annual power needs.

Whether these planners are clairvoyants or dreamers is a question that will have to wait to be answered, of course, but as writer and futurist Arthur C. Clarke has pointed out, humans have always tended to overestimate what can be accomplished in the near future and to underestimate grossly what will be achieved in the longer term. It is not outlandish to imagine that fifty years from now nonpolluting power will be beamed to the earth from space, that virtually all earth-based communications will be relayed by orbiting satellites, and that dozens or even hundreds of people will live and work in several orbiting villages, devoted to the performance of a variety of proven and successful space tasks.

It also seems reasonable to conjecture that by the middle of the next century, a permanently occupied outpost will be based on the moon, and that minerals mined there will be used in the manufacture of satellites, space stations, and planetary probes. Several year-long

Twenty-three years after the first humans orbited the earth, astronaut Bob Stewart operates the manned maneuvering unit on its first experimental mission, flying free of his spacecraft, a human satellite entering space.

manned missions to Mars may have been successfully completed, and plans may be under way for a new generation of explorers to journey to the four Galilean moons of Jupiter.

Space, the realm that encompasses the whole universe, is incredibly close at hand. From any point on the earth, space is only about eighty miles away. The earth itself is as much a part of space as is the moon, or meteors, or the Milky Way, yet without the aid of rockets that can hurl us away from our planet's hold, we can only dimly peer beyond the atmosphere; we cannot venture away from home.

The fundamental reasons for entering space are rooted in the nature of our species. We are the inquiring species, the creatures whose survival depends on knowledge and its use. The promise of space is the promise of tangible solutions to environmental and resource problems, of improved communication among the peoples of the earth, of new technologies that hold the potential to help us lead more productive lives.

The entire course of the space program—from the journeys into orbit of the Mercury capsules, and the flights of the Apollo command and lunar modules, to the exploration of the planets by robotic spacecraft and the advent of the versatile, reusable shuttle orbiters—has allowed us to become less theoretical about the nature of the earth, its neighboring spheres in the solar system, and the emptiness that surrounds us all. We still know little about the cosmos; space is still a mysterious frontier, but because we have *been there,* because we have begun to enter space, we know far more than we could have ever known had we chosen not to go. Space will forever lure us outward because it is a realm without boundaries, without limitations, without an end to the promise of understanding.

SUGGESTED READINGS

Arthur C. Clark. *The Promise of Space*. New York: Harper & Row, 1968.

Michael Collins. *Carrying the Fire*. New York: Farrar, Straus & Giroux, 1974.

Henry S. F. Cooper, Jr. *A House in Space*. New York: Holt, Rinehart & Winston, 1976.

Oriana Fallaci. *If the Sun Dies*. New York: Atheneum, 1965.

Timothy Ferris. *Galaxies*. New York: Stewart, Tabori & Chang, 1982.

Bruce Murray, ed. *The Planets*. San Francisco: W. H. Freeman, 1983.

Gerard K. O'Neill. *The High Frontier: Human Colonies in Space*. New York, William Morrow, 1977.

Carl Sagan. *Cosmos*. New York: Random House, 1980.

Tom Wolfe. *The Right Stuff*. New York: Farrar, Straus & Giroux, 1979.

Additional Photo Credits

Back Flap: Hope Lieberstein/James Long Associates
22–23: James Long/James Long Associates
37: Jim Rousseau/James Long Associates
38: Fernando Cruz/James Long Associates
42: Robert N. Cerri/James Long Associates
47: Fernando Cruz/James Long Associates
49: Jim Rousseau/James Long Associates
54: Jim Rousseau/James Long Associates
57, left: Harvey W. Olsen/James Long Associates
57, right: Thom Baur/James Long Associates
203: James Long/James Long Associates

INDEX

(Numerals in *italics* indicate a photograph)

The text was set in Cloister by
TGA Communications, Inc.,
New York, New York.

The book was printed and bound by
Dai Nippon Printing Co., Ltd., Tokyo, Japan.